UNIT

CCEA AS 2

Biology

Organisms and Biodiversity

John Campton

Philip Allan Updates, an imprint of Hodder Education, an Hachette UK company, Market Place, Deddington, Oxfordshire OX15 0SE

Orders
Bookpoint Ltd, 130 Milton Park, Abingdon, Oxfordshire, OX14 4SB
tel: 01235 827720
fax: 01235 400454
e-mail: uk.orders@bookpoint.co.uk
Lines are open 9.00 a.m.–5.00 p.m., Monday to Saturday, with a 24-hour message answering service. You can also order through the Philip Allan Updates website: www.philipallan.co.uk

© Philip Allan Updates 2010

ISBN 978-0-340-99194-7

First printed 2010
Impression number 5 4 3 2
Year 2014 2013 2012 2011 2010

This guide has been written specifically to support students preparing for the CCEA AS Biology Unit 2 examination. The content has been neither approved nor endorsed by CCEA and remains the sole responsibility of the author.

Printed by MPG Books, Bodmin

Hachette UK's policy is to use papers that are natural, renewable and recyclable products and made from wood grown in sustainable forests. The logging and manufacturing processes are expected to conform to the environmental regulations of the country of origin.

Contents

Introduction

■ ■ ■

Content Guidance

■ ■ ■

Questions and Answers

Introduction

About this guide

The aim of this guide is to help you prepare for the AS Unit 2 examination for CCEA biology. It also offers support to students studying A2 biology, since topics at A2 rely on an understanding of AS material.

This guide has three sections:
- **Introduction** — this provides guidance on the CCEA specification and offers suggestions on improving your study/revision skills and examination technique.
- **Content Guidance** — this summarises the specification content of AS Unit 2.
- **Questions and Answers** — this provides two exemplar papers for you to try. There are answers written by two candidates and examiner's comments on the candidates' performances and how they might have been improved.

Try to adopt the suggestions given in this introduction about how to study. This will affect your performance throughout the course — it takes time to learn how to study at this level.

The Content Guidance should be used as a study aid as you meet each topic, for end-of-topic tests, and during your final revision. There are seven topics and at the end of each topic there is a list of the practical work with which you are expected to be familiar.

The Questions and Answers section will be particularly useful during your final revision. It presents a range of question styles that you will encounter in the AS Unit 2 exam, and the candidates' answers and examiner's comments will help with your examination technique.

The specification

You should have your own copy of the CCEA biology specification. This is available from **www.ccea.org.uk**.

AS biology

The AS course is made up of three units. Units 1 and 2 are assessed by examination papers. Unit 3 is assessed internally and has a lower weighting.

AS Unit	Title	Weighting	Availability
1	Molecules and Cells	40% of AS	January and summer
2	Organisms and Biodiversity	40% of AS	January and summer
3	Assessment of Practical Skills	20% of AS	Summer

This guide covers AS Unit 2; the first book in the series covers AS Unit 1. AS Unit 3, is based on the practical work (i.e. coursework) that you will do in biology classes

Assessment objectives

AS biology is not just about remembering facts. Examiners need to assess your *skills* — your ability to do things and work things out.

Examinations in AS biology test three different assessment objectives (AOs). AO1 is about remembering the biological facts and concepts covered by the unit. AO2 is about being able to use the facts and concepts in new situations. AO3 is called How Biology Works. It emphasises that biology, as a science, develops through testing hypotheses. It involves an understanding of practical procedures and the analysis of results to determine whether a hypothesis has been supported or disproved.

The following table gives a breakdown of the approximate number of marks awarded to each AO in the examination.

Assessment objective	Description	Marks
AO1	Knowledge and understanding of biology and of How Biology Works	32 (42.5%)
AO2	Application of knowledge and understanding of biology and of How Biology Works	32 (42.5%)
AO3	How Biology Works	11 (15%)

The AS Unit 2 paper

The AS Unit 2 examination lasts 1 hour 30 minutes and is worth 75 marks. The paper consists of about nine questions; the mark allocation per question is from 3 to 15 marks. There are two sections. In Section A all the questions are structured; in Section B there is a single question, which may be presented in several parts, and which should be answered in continuous prose.

Questions towards the start of the paper and the initial parts of questions tend to assess straightforward 'knowledge and understanding' (AO1). There will also be questions that present information in new contexts and may test your skills in analysing and evaluating data (AO2).

AO3 may be assessed by questions that ask you to describe experimental procedures. You may also be asked to demonstrate graphical or drawing skills, since these are important aspects of biology. Graphs are used to illustrate quantities; drawings illustrate the qualities of a feature. There are other skills that you may be asked to demonstrate — for example, organisation of raw data into a table.

You are expected to use good English and accurate scientific terminology in all your answers. Preparing a glossary of terms used in each topic should aid this. Quality of written communication is assessed throughout the paper and is specifically awarded a maximum of 2 marks in Section B.

Study and revision skills

Students who achieve good grades have good study strategies. This section of the Introduction provides advice and guidance on how these might be achieved.

Revision is an ongoing process

Revision should not be just something that you do before an exam. It should be continual throughout the course. Work consistently and complete each task as the teacher sets it. Use study periods in school to develop your understanding, *not* for homework. Study thoroughly for tests. Try to find time at the weekend to go over that week's work. If you keep going over topics then you won't panic with the intensive revision required at exam time.

Active learning is best

Just reading through your notes or a chapter of a textbook is not a particularly effective way to revise. You simply learn material to later forget it. In order to develop a deeper understanding, you have to use more than just the 'reading' centre of the brain — you should make your brain *do* something with the material. It is for this reason that you must write your own notes. You should also try to do things in different ways, for example:

- a series of bullet points
- a flow diagram
- an annotated diagram
- a spider diagram
- a prose account

Particular methods are appropriate to the topic. A spider diagram on the heart would include reference to its structure, the wave of excitation, the different phases of the cardiac cycle, pressure changes and the operation of the valves. Annotated graphs would be a good way to revise the pressure changes during a cardiac cycle. Compiling a glossary of terms to do with heart action will improve your understanding of how each term is defined. An essay on the cardiac cycle will test your understanding of the entire topic and give you practice for the Section B question.

Developing your understanding

You should learn to develop your understanding so that you can apply it. Use different texts and the internet. These will present the information in different ways, so that your brain can perceive it from different perspectives. It is only when you understand a topic fully that you will be able to deal with questions that set the topic in a new context. If you have problems with a difficult concept, ask your teacher to explain it in a different way. Teachers are happy to help, as long as you have worked at the topic yourself. Working with another student may also help. Remember too that AS biology is a step up from GCSE. There will be difficult concepts and so you must persevere.

Organise your notes

You will have accumulated a large quantity of notes from your teacher and, more importantly, those that you have made yourself. You should organise your notes under headings and sub-headings and construct a summary of the key points. The Content Guidance section will help you do this. It is essential that you keep this information in an organised manner. Gather all your notes, divide them into sections on each topic and keep them in a file for each unit. This will make it easier both when you make additional notes and during your final revision.

Planning your revision

While revision is an ongoing process, you will have to undertake intensive revision in the weeks before the exam. Make out a revision schedule taking into account all the topics you have to cover and the time available. This can seem alarming, but if you break down the total amount into smaller portions then it will become achievable. Make sure that each part of the unit gets its fair share of your attention and allocate more time to difficult areas. Try to leave time in your schedule to practise past questions.

Keeping your concentration

Some students 'get lost' in their work and can concentrate for hours. Some find it difficult to concentrate for longer than an hour. You should do whatever is appropriate for you, but try to revise in a quiet place with no distractions. You should take breaks, which could be organised as rewards — for example, a favourite television programme.

Vary your revision and keep it active

In the final revision it is easy to revert to just reading through your notes. Try to keep it active by summarising what you have. This keeps your brain processing the information. Try to vary what you do. You should test your understanding by practising past questions, such as in the exemplar papers in the Questions and Answers section of this guide. You should practise skills such as calculations and also write essays on the various topics in the unit. Scanning can be an effective revision technique: read the first sentence of each paragraph (the rest of the paragraph generally only provides elaboration); read sentences with key words (often in bold); and, in particular, study diagrams, since these often summarise important information. You might also find the website at **http://highered.mcgraw-hill.com/sites/dl/free/0072437316/120060/ravenanimation.html** useful for animations of many key processes.

The examination

Before the day of the examination it is important that you are well prepared. You should have all the implements that you will need: two black pens and two pencils; your calculator plus spare batteries; a ruler and an eraser. Try to get a good night's sleep. When you enter the exam hall, tell yourself that you really understand this unit — *be positive*. You will always know more than you think you do. A few nerves are good and will help you stay more alert during the exam.

Time

You have 90 minutes to answer questions worth a total of 75 marks. That gives you over 1 minute per mark, so there is some preparation time and you should have time at the end to go over your answers. At the start of the exam it can be beneficial to spend some time looking through the paper and scanning the questions, especially the question in Section B. This will allow you to think of relevant points while answering other questions. If you get stuck, make a note of the question number and move on — you can come back later. You are advised to spend 20 minutes on Section B. Try to keep to this, but leave time for writing a plan of the information you want to include. At the end of the exam go over the paper, including those parts about which you were unsure, correcting any mistakes or filling in gaps. It is also beneficial to double-check calculations to make sure that you haven't made any silly mistakes.

Read the questions carefully

This sounds obvious, but you can lose many marks by not doing so. There are two aspects:

- You must understand the command terms used in the question, i.e. the word at the start of the question. Appendix 1 of the specification and a guide in the biology microsite on **www.ccea.org.uk** explain these terms. The two terms that are most commonly misinterpreted are 'describe' and 'explain'. *Describe* requires you to provide an accurate account of the main points. You do not need to offer an explanation. You may be asked to describe something on the paper, such as a graph. You may be asked to describe a situation or a process — for example by writing out the sequence of events in the loss of water by transpiration. When describing what is shown in a graph or a table, marks can often be gained by making appropriate reference to the data — for example the point at which a particular change takes place. *Explain* requires you to provide reasons for why or how something is happening. A description is not required. 'Explain how...' means that you should show the way something happens. 'Explain why...' is asking you to give reasons for something, such as an event or outcome.
- The stem of a question may provide information needed to answer the question. Don't ignore this information. You might wish to highlight or underline these parts

of the question stem. Think about how this information can help you to construct or focus on a relevant answer.

Depth and length of answer

The examiners give you guidance about how much you need to write:

- **The number of marks**. In general, the number of marks indicates the number of points that you should provide — for a question worth 4 marks you need to give more points than for one worth 2 marks. For questions testing straightforward recall, you may need to provide more points than there are marks — for example, three points for 2 marks. However, this should be obvious from the number of spaces. The number of marks is the most important guideline with respect to the depth of answer required.
- **The number of lines**. In general, there will be two lines for each marking point. However, examiners expect you to keep your answer relevant and precise, so occasionally there may be only one line. In Section B, the number of lines allocated is generous — a good answer will not necessarily use them all.
- **The recommended time**. You are advised to spend 20 minutes on Section B. Try to keep to this. It is possible to write a longer 'essay' but you would be providing more points than there are marks available.

Quality of written communication (QWC)

The ability to organise thoughts, express ideas clearly and to make use of the appropriate terminology is an important aspect of biology. In Section A questions, credit may be restricted if communication is unclear. Where QWC is assessed in Section A, the mark schemes for questions will contain specific statements. These statements will relate to the clear expression of concepts (e.g. 'transpiration creates a more negative pressure in the leaf resulting in a water potential gradient along which water moves', rather than 'transpiration sucks water up'), correct biological terminology (e.g. use of the term 'partial pressure' with respect to oxygen levels, rather than 'concentration') and correct spelling where there are close alternatives (e.g. gene and genus). Note that mistakes in spelling are not generally penalised as long as the examiner knows what you are trying to say. With respect to clarity in answers, a common problem is to use the word 'it' in such a way that the examiner can't be certain what 'it' refers to — so try to avoid using 'it' in answers. In Section B there is a maximum of 2 marks available for QWC. The examiners want to see well-linked sentences that present relationships and do not just list features.

Content
Guidance

This section summarises what you need to know and understand for the AS Unit 2 examination paper. It is divided into seven topics:

- **Principles of exchange and transport** — an overview of the factors that influence the absorption and exchange of metabolites, and their movement within an organism
- **Gas exchange in flowering plants and mammals** — the exchange of oxygen and carbon dioxide during respiration and photosynthesis in flowering plants, and in the lungs of mammals
- **Transport and transpiration in flowering plants** — xylem and phloem as vascular tissues, transpiration and the movement of water, and the translocation of organic substances
- **Blood transport system in mammals** — the structure and function of the heart, blood vessels, blood, haemoglobin and the carriage of oxygen, and cardiovascular disease
- **Diversity of life** — the classification of organisms and biodiversity
- **Adaptation of organisms** — the adaptations of organisms to their environment and influences on their distribution, and the role of natural selection in maintaining their adaptability
- **Human impact on biodiversity** — how local biodiversity has been adversely affected and strategies used to improve the situation

At various points within the section there are examiner's tips. These offer guidance on how to avoid difficulties that often occur in examinations.

At the end of most topics there is a list of practical work with which you should be familiar.

Principles of exchange and transport

Living cells require certain substances to maintain their metabolic processes. These substances include:

- oxygen for aerobic respiration
- glucose and fatty acids as substrates for respiratory metabolism
- fatty acids for the synthesis of phospholipids required for the production of membranes
- amino acids for the synthesis of proteins (e.g. enzymes)
- ions to maintain water potential, as enzyme cofactors and for other aspects of metabolism (e.g. in plant cells, nitrate ions to provide nitrogen for amino acid synthesis and phosphate for the synthesis of phospholipids)
- water as a solvent

These substances are either synthesised within the cell (in which case other substances are usually required), or released from storage molecules within the organism (and perhaps transported through the organism), or obtained from the environment.

All living cells need to remove the toxic by-products of their metabolic processes. These include:

- carbon dioxide from aerobic respiration (in animal cells and micro-organisms, and from plant cells not actively carrying out photosynthesis)
- oxygen from photosynthesis (in green plant cells)
- urea and ammonia from excess amino acids (in animals, since plants generally only synthesise the amino acids they require)

To summarise, animal tissues need to obtain:

- oxygen from the air (or from water if they are aquatic)
- glucose, fatty acids and amino acids from food
- water

Carbon dioxide and nitrogenous waste have to be removed from animal tissues.

The tissues of flowering plants need:

- oxygen from the air, especially at night
- carbon dioxide from the air during the day
- inorganic ions (which, together with sugars from photosynthesis, produce amino acids etc.) from the soil solution
- water from the soil solution

Depending on the time of day, either carbon dioxide or oxygen is removed from plant tissues.

Thus, organisms need to absorb or exchange substances with their environment and transport these substances within the organism.

The exchange of substances

There are a number of factors that influence the absorption or exchange of substances.

The exchange of substances occurs only at **moist permeable surfaces**. Aquatic organisms, organisms that live in the soil and even some terrestrial organisms, such as amphibians, all have moist and so permeable body surfaces. This is not the case with terrestrial organisms such as mammals and flowering plants. To prevent water loss by evaporation, mammals possess an impermeable surface (skin with a cornified layer onto which an oily substance is secreted) while the aerial parts of flowering plants are covered in a waxy cuticle secreted by the epidermis. This means that mammals and flowering plants must have specialised absorptive or exchange surfaces.

The need for specialised absorptive or exchange surfaces depends on the size and shape of the organism because both these factors affect an organism's **surface area-to-volume ratio**. The surface area is represented by the total number of cells in direct contact with the surrounding environment. The volume is the total three-dimensional space occupied by metabolically active tissues. The absorptive surface area is a measure of the rate of supply of metabolites to tissues. The volume of the organism is a measure of its demand for metabolites. An organism must be capable of taking up sufficient material to satisfy its needs. Therefore, the surface area-to-volume ratio is critical — it must be sufficiently large.

The influence of size on the surface area-to-volume ratio is shown graphically in Figure 1, in which cubes of different dimensions are used as the example.

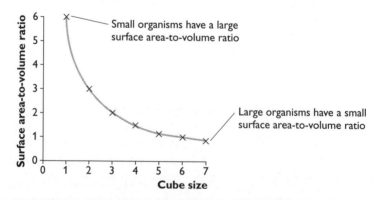

Figure 1 The relationship between size and surface area-to-volume ratio

While this relationship has been calculated for cubes, the principle is true for all regular shapes. *Small* organisms have a *large* surface area-to-volume ratio. The external surface of a small organism can be used as a gas exchange surface because the large surface area is able to supply sufficient oxygen to the small volume. For example, an earthworm is small enough for its body surface to be used for gas exchange.

A larger organism has a small surface area compared with its large volume, i.e. a small surface area-to-volume ratio. The large volume creates a demand for oxygen which the small surface area is unable to supply. Therefore, large organisms need specialised permeable surfaces whereby the absorption or exchange area is increased to satisfy the needs of the organism.

The rate at which an organism requires substances depends on its metabolic rate. An organism with a high metabolic rate has a high oxygen requirement and, therefore, possesses specialised, large, gas exchange surfaces. This can be illustrated by comparing the mouse and the frog, which are organisms of similar size. The mouse has a metabolic rate and oxygen consumption approximately ten times greater than that of a frog, and has a proportionately bigger gas exchange surface. The frog's lungs are simple sacs, while the mouse has spongy lungs consisting of millions of microscopic alveoli. The large surface area is essential for the high rate of oxygen uptake required for the high metabolic rate of the mouse.

To summarise, an organism requires a specialised absorptive surface if it is terrestrial (with an impermeable surface), large (with a small surface area-to-volume ratio) or has a high metabolic rate. Methods of increasing the area of an absorptive surface include:
- evagination (outfolding) of the surface
- invagination (infolding) of the surface
- flattening of the organism

Organisms with a flattened shape have a large surface area-to-volume ratio. A cube of $1 \times 1 \times 1$ arbitrary units has a surface area-to-volume ratio of 6. If this cube is flattened to dimensions of $0.1 \times 10 \times 1$ arbitrary units, it has the same volume but the surface area-to-volume ratio is 22.2, an increase of nearly four-fold. Flattening not only increases the surface area-to-volume ratio, it also decreases the distance over which substances have to be moved.

> **Tip** To ensure that your learning is active you should go over the calculation of surface area-to-volume ratios shown in Figure 1 and described in the previous paragraph. Remember that the examination will require you to carry out at least one calculation, so practice is a good idea. The ratio is calculated as the proportion of one value to another. Apart from ratio, you might also be asked to calculate magnification (see Student Unit Guide AS 1), percentage, percentage change and rate.
>
> Percentage means out of 100. It is calculated by dividing a value by the total and multiplying by 100. Percentages are calculated in J-tube analysis (determination of the percentage O_2 and percentage CO_2 content of air samples).
>
> To calculate a percentage change, divide the difference by the initial amount and multiply by 100. You should be aware that:

- a negative change represents a decrease
- a positive change means that there is an increase
- a 100% increase equates to a doubling in the quantity
- a 1000% increase is a ten-fold increase

The rate of a process is calculated as the change per unit time.

Table 1 shows some important examples of efficient absorptive surfaces in flowering plants and mammals.

Table 1 Absorptive surfaces in flowering plants and mammals

Absorptive surface	Structure	Function
Leaf mesophyll Palisade mesophyll Spongy mesophyll Air space system	The leaf is a flattened structure (its thinness ensures a short diffusion distance) with a tightly packed upper palisade mesophyll layer and a loosely packed lower spongy mesophyll layer	Wide expanse of palisade tissue is efficient at trapping light; the loose arrangement of the spongy layer provides an air space system through the leaf and creates a huge surface for gas exchange
Root hairs Epidermis Root hair	Tubular extensions of the epidermal cells of the young root	Increase greatly the surface area of the root for the uptake of oxygen, water and ions
Alveoli Terminal bronchiole Alveolar duct Alveoli	Small (diameter 0.2 mm) sacs, occurring in clusters and in vast numbers within the mammalian lung; in human lungs there are 700 million, providing a total surface area of 70 m^2	Huge, moist surface area provides for efficient gas exchange; the alveolar walls are thin (0.1–1.0 µm), so the diffusion distance is short
Capillaries Capillary network	Small (diameter 5–10 µm), thin-walled blood vessels, with a total length of 100000 km and surface area of 1000 m^2 in the human body	Extensive networks throughout the body represent a huge surface area for the exchange of molecules between the blood and the body tissues; the number and distribution of capillaries is such that no cell is further away than 50 µm from a capillary, i.e. the diffusion distance is minute

Absorptive surface	Structure	Function
Erythrocytes (red blood cells) Top view Side view	Small (diameter 8 µm) flexible biconcave discs, flattened and depressed in the centre, with a dumbbell-shaped cross section	The biconcave disc shape greatly increases the surface area-to-volume ratio for efficient uptake of oxygen; the thinness of the cell, particularly where it is depressed in the centre, allows oxygen to diffuse to all the haemoglobin packed into the cell

The transport of substances

The molecules in gases and liquids move constantly and at random. If there is a difference in the concentration of molecules within an area, a net movement of the molecules occurs, resulting in the molecules becoming evenly distributed. This is illustrated in Figure 2.

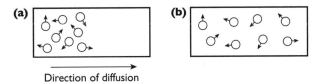

Direction of diffusion

Figure 2 Random movement of molecules causes them to become dispersed from an area of high concentration (a) until they are evenly distributed (b)

Such movements are known as **diffusion**. Diffusion can take place across surfaces as long as these are moist, and permeable to the substances.

Diffusion is a sufficiently effective mechanism for the movement of substances across thin surfaces and within thin organisms. For example, a flatworm (an animal with a simple body plan, small, tapering and flattened — about 2 cm long and 2 mm thick) absorbs oxygen at its moist body surface and this oxygen diffuses to all body tissues since these are within easy reach of the surface. However, diffusion alone does not suffice for the movement of substances within large organisms.

Transport of substances in larger organisms occurs by mass flow (Figure 3). In mass flow, unlike diffusion, all the molecules are swept along in the same direction.

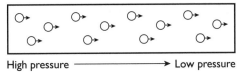

High pressure ⟶ Low pressure

Figure 3 The mass flow of molecules

Mass flow is brought about by a pressure difference. Mass flow systems include the xylem and phloem systems of flowering plants and the breathing (ventilation) and blood circulatory systems of mammals. Different mass flow systems have different means of generating a pressure difference (Table 2).

Table 2 Examples of mass flow systems in flowering plants and mammals

Mass flow system	Method of generating a pressure difference	Function of the mass flow system
Xylem system	Tension (negative pressure) in the leaf xylem generated by the transpirational loss of water from the leaves	One-way transport of water and ions from roots to leaves in a flowering plant
Phloem system	Movement is driven by energy from the plant	Two-way flow of organic solutes (e.g. sucrose) in a flowering plant
Breathing (ventilation) system	Pressures in the thorax are alternately decreased (inducing inhalation) and increased (inducing exhalation)	Ventilation of the mammalian lungs, whereby air is alternately drawn in and forced out
Blood circulatory system	High pressure is generated by the muscular heart	Circulation of blood carrying oxygen, glucose, amino acids, fats, carbon dioxide, urea and other substances in a mammal

Tip There is an overlap between the principles of exchange and of transport. For example, the flatness of an organism (or of a cell) increases its surface area-to-volume ratio and, therefore, its effectiveness in absorption. However, it also decreases the distance of the diffusion path, which is probably more important. An earthworm, though bigger than a flatworm, has a sufficiently large surface area-to-volume ratio to satisfy its gas exchange requirements, but needs an internal transport (blood circulatory) system, since the diffusion distance is too great. The flatworm has no need for an internal transport system, not only because it is thin, but also because its digestive system branches throughout its body so that food is absorbed in close proximity to all tissues (see Figure 4).

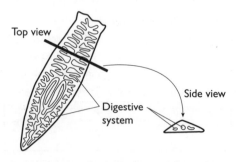

Figure 4 The flatworm: an animal with no need for a blood circulatory system

Note that reference has been made to flatworms, earthworms and even frogs, none of which is specified in the content of the unit. This is simply because you are required not only to have knowledge of the principles of exchange and transport, but to be able to apply your understanding in unfamiliar situations. Therefore, it is better to learn these principles through appropriate examples.

Gas exchange in flowering plants and mammals

The biological processes of respiration and photosynthesis require one gas which they exchange for another. The surface over which gas exchange takes place must:
- be **permeable** to oxygen and carbon dioxide
- be **moist**, since gases must dissolve in water before diffusing into tissue cells
- have a sufficiently **large surface area** to satisfy metabolic requirements

Gas exchange in flowering plants

There are two processes in flowering plants that involve gas exchange: **respiration** and **photosynthesis**. Respiration takes place in all tissues, all the time. Photosynthesis takes place only in green tissues (i.e. those containing chlorophyll), and only during the daylight hours. Indeed, the rate of photosynthesis depends on the light intensity. So, the maximum rate of photosynthesis occurs when the light intensity is highest, such as at midday. At this point, the rate of photosynthesis greatly exceeds the rate of respiration and so there is a net production of oxygen. At a specific low light intensity (during dawn and dusk) the rate of photosynthesis equals the rate of respiration and so the net exchange of oxygen is zero. This is known as the **compensation point**, since the rate of oxygen production (in photosynthesis) is balanced by the rate of oxygen consumption (in respiration).

These changes in oxygen use and release by a flowering plant are shown in Figure 5.

Respiratory gas exchange in flowering plants
The roots of plants use energy in processes such as cell division (growth) and the active transport of ions from the soil solution. In the growth region, the epidermal cells possess **root hairs**, which increase the surface area-to-volume ratio. In soil that is not waterlogged, **root hairs** are surrounded by air spaces between the particles of soil. Diffusion of respiratory gases occurs through the cell wall and cell membrane of these root hairs (Figure 6).

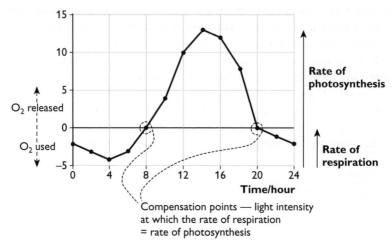

Figure 5 Changes in oxygen used and released by a plant over a 24-hour period

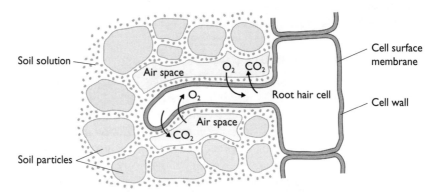

Figure 6 Gas exchange in a root hair cell

In stems, especially those of woody plants, the most active cells are those under the surface. Although the stem's outer covering is waterproofed to reduce water loss to the air (and so is impermeable to gases) there are small pores to allow oxygen in and carbon dioxide out.

Plants lack specialised respiratory surfaces and yet can be very large. This is because they lack tissues with a high energy demand and so have low respiration rates.

Photosynthetic gas exchange in flowering plants

The **leaf** is the major photosynthetic organ in a flowering plant. The leaf needs a specialised gas exchange surface because:

- a high rate of photosynthesis is generated
- the concentration of carbon dioxide in the air is low (390 parts per million or 0.039%)

content guidance

The leaf epidermis, particularly the lower epidermis, possesses **guard cells** that control the opening and closure of **stomata**. The stomata open during the hours of daylight. When open, air, containing carbon dioxide, diffuses into and out of the leaf mesophyll. The carbon dioxide diffuses through an **air space system** provided by the **spongy mesophyll**. Having diffused through the air space system, the carbon dioxide is absorbed by the mesophyll cells (in which the carbon dioxide concentration is low as it is used in photosynthesis). It is this moist mesophyll surface that represents the gas exchange surface (see Figure 7). Since the leaf is *broad and thin*, there is a large surface area and the diffusion distance for gases is short.

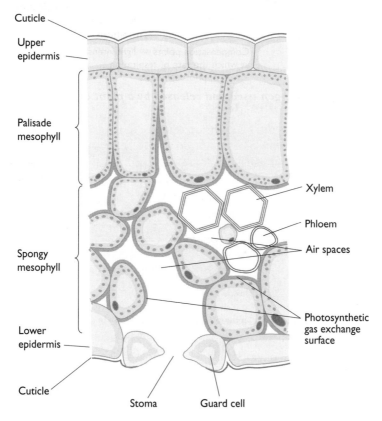

Figure 7 The photosynthetic gas exchange surface of a leaf

Oxygen produced in photosynthesis diffuses out of the cells into the air space and then out through the open stomata. Some oxygen is, of course, used up in respiration.

Plants living in water: hydrophytes

There is much less oxygen dissolved in water than there is oxygen in air, so the stems and leaves of aquatic flowering plants (hydrophytes) have adaptations to facilitate the uptake and movement of oxygen and carbon dioxide (see Figure 8).

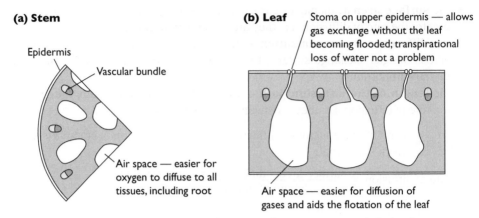

(a) Stem

Epidermis

Vascular bundle

Air space — easier for oxygen to diffuse to all tissues, including root

(b) Leaf

Stoma on upper epidermis — allows gas exchange without the leaf becoming flooded; transpirational loss of water not a problem

Air space — easier for diffusion of gases and aids the flotation of the leaf

Figure 8 Adaptations of a hydrophyte (a) stem and (b) leaf

Gas exchange in mammals

Gas exchange in cells occurs by diffusion (Figure 9). In cells respiring aerobically, oxygen is used and carbon dioxide is produced. This affects the concentration gradients and so oxygen diffuses into the cell and carbon dioxide diffuses out.

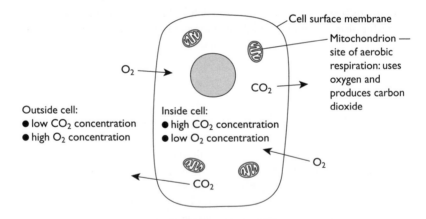

Cell surface membrane

Mitochondrion — site of aerobic respiration: uses oxygen and produces carbon dioxide

O_2

CO_2

Outside cell:
● low CO_2 concentration
● high O_2 concentration

Inside cell:
● high CO_2 concentration
● low O_2 concentration

O_2

CO_2

Figure 9 Gas exchange in a respiring cell

The movement of oxygen to cells, and of carbon dioxide diffusing out from the body cells, involves four stages:
- diffusion of gases between respiring cells and the blood
- transport of gases in the blood
- diffusion of gases across the gas exchange surface between the alveolar air and the blood
- ventilation of the lungs with fresh air

The **rate of diffusion** across the gas exchange surface depends on:

- the **surface area** available for gas exchange — the greater the surface area, the greater the rate of diffusion, i.e. they are directly proportional
- the **difference in concentration** — the greater the gradient in concentration across the gas exchange surface, the greater the rate of diffusion, i.e. they are directly proportional
- the **length of the diffusion path** — the shorter the diffusion path, the greater the rate of diffusion, i.e. they are inversely proportional

The relationship of these factors in affecting the rate of diffusion is summarised by Fick's law, which states that:

$$\text{rate of diffusion} \propto \frac{\text{surface area} \times \text{difference in concentration}}{\text{length of diffusion pathway}}$$

Gas exchange in the lungs is particularly efficient. The walls of the alveoli provide a large surface area and consist of thin cells (as do the capillary walls); concentration gradients are maintained by the ventilation of the lungs and the circulation of blood.

The structure of the lungs

Air is breathed through the nostrils or mouth, and enters or leaves the lungs via the **trachea**. The lungs are situated within the **thorax** (also known as the **thoracic cavity**). The trachea branches into two **bronchi** (singular **bronchus**) which further branch into a series of ever-finer bronchioles forming a **bronchial tree**. Each **terminal bronchiole** leads to a cluster of **alveoli**, an **alveolar duct** connecting with each **alveolus**. Each individual alveolus is tightly wrapped in blood capillaries and it is here that gas exchange takes place. The structure of the lung system is shown in Figure 10.

Ventilation of the lungs

Changing the volume of the thoracic cavity changes the air pressure inside the lungs. Air moves from a region of high pressure to a region of low pressure. When the volume of the thorax is increased, the pressure in the lungs is decreased, becoming lower than atmospheric pressure so that air moves into the lungs (inhalation or inspiration). When the volume of the thorax is decreased, the pressure in the lungs is increased, becoming higher than atmospheric pressure so that air moves out of the lungs (exhalation or expiration). The mechanisms of **inhalation** and **exhalation**, during normal breathing, are summarised in Figure 11.

During forced exhalation, such as during hard exercise and coughing, the volume of the thoracic cavity is further reduced by the contraction of the internal intercostal muscles depressing the rib cage.

Alveolar structure and gas exchange

In each human lung there are about 350 million alveoli, each of which is about 200 μm in diameter. Their total surface area is about 70 m². Each alveolus is lined with a single layer of squamous (pavement) epithelial cells only 0.05 μm to 0.3 μm thick. The inner surface of each alveolus is moist. Around each alveolus is a network of blood capillaries so narrow (7–10 μm) that, in order to squeeze through, erythrocytes (red blood cells) are flattened against the capillary walls. These capillaries also have walls one cell thick and consisting of a squamous epithelium (0.04–0.2 µm thick).

Gases can diffuse across the moist, permeable walls of the alveoli. Diffusion of gases between the alveoli and the blood is rapid because:
- concentration gradients are maintained by the ventilation of the lungs and the blood flow through the pulmonary capillaries
- the alveoli and pulmonary capillaries have very large surface areas
- the walls of both alveoli and capillaries are thin and therefore the distance over which diffusion takes place is very short — in fact the distance between alveolar air and the erythrocytes averages only about 0.5 µm.

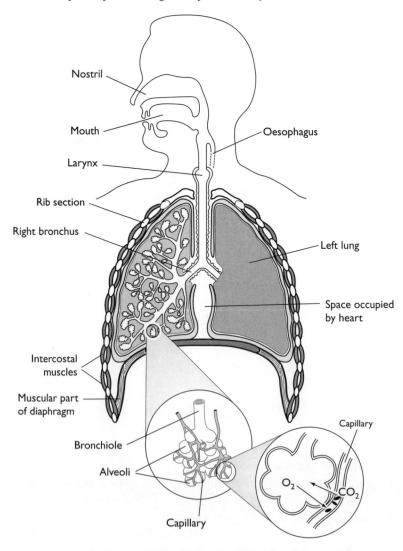

Figure 10 The structure of the lung system

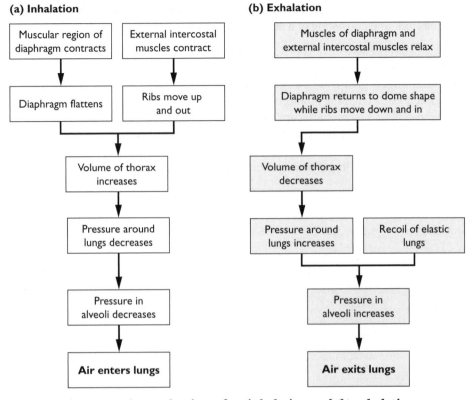

Figure 11 The mechanism of (a) inhalation and (b) exhalation

The structure of an alveolus and associated capillaries is shown in Figure 12 (though the walls, for clarity, have been drawn disproportionately thick).

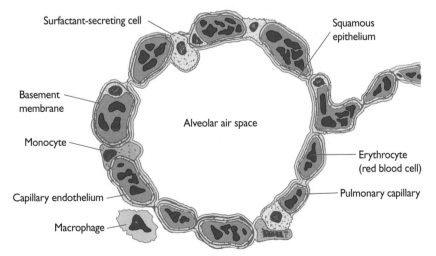

Figure 12 The structure of an alveolus and associated pulmonary capillaries

Since the alveolar surface is situated deep inside the body, evaporation of water from its moist surface is reduced to a minimum. Other types of cell are present in the alveolar wall. **Surfactant-secreting cells** (septal cells) produce a detergent-like substance that reduces the surface tension in the fluid coating the alveoli, and without which the alveoli would collapse due to the cohesive forces between the water molecules lining the air sacs. **Macrophages** (derived from **monocytes**, a type of white blood cell) protect the lungs from a broad spectrum of microbes and particles by ingesting them through phagocytosis. Elastic fibres are also associated with the alveolar walls, and the elastic recoil of the alveoli helps to force air out during exhalation.

Smoking and lung disease

Tobacco smoke contains thousands of toxic substances, many of which are collectively known as tar. Therefore, it is not surprising that smoking has been linked to a number of lung diseases. It is common for smokers to get a combination of **chronic bronchitis** and **emphysema**, which is called chronic obstructive pulmonary disease (COPD).

Tar brings about an inflammatory response in which the airways narrow and excessive amounts of mucus are produced. Furthermore, tar paralyses the cilia that sweep mucus and bacteria away from the lungs, so pathogens and mucus build up. This leads to phlegm production, coughing and breathlessness — the symptoms of chronic bronchitis. The inability to clear mucus and bacteria results in an increased susceptibility to chest infections, including pneumonia.

In emphysema, the inflammatory response to smoke inhalation leads to the breakdown of the walls of the alveoli. This reduces the area available for gas exchange so it becomes difficult to get enough oxygen. There is also a loss of elastic fibres in the alveolar walls. Therefore, exhalation becomes more difficult because the ability of the alveoli to recoil following inhalation is reduced.

Tobacco smoke contains many **carcinogens** — substances that may induce **cancer** — of which tar is the most important. Carcinogens may damage the DNA in the cells lining the bronchial tubes. Cells with DNA damage may divide in a modified and uncontrolled way producing a mass of unspecialised cells known as a tumour. A cancerous or malignant tumour may, over time, spread to invade other tissues.

Practical work

Use a J-tube to analyse air samples:
- determine the percentage oxygen consumption and percentage carbon dioxide consumption of samples of inspired and expired air

Understand the use of a simple respirometer:
- measure oxygen consumption (with potassium hydroxide present)
- measure the net difference between carbon dioxide production and oxygen consumption (in the absence of potassium hydroxide) and so determine carbon dioxide production

Understand the use of the Audus apparatus (photosynthometer):
- the effect of light intensity on the photosynthetic production of oxygen

Demonstrate a compensation point:
- use of bicarbonate indicator

Transport and transpiration in flowering plants

Vascular tissues

There are two transport systems in flowering plants:
- **xylem** for water and inorganic ions
- **phloem** for organic molecules, such as sucrose and amino acids

These systems are contained within the plant's **vascular tissues**. Vascular tissues form a central **stele** (vascular cylinder) in the root, but peripheral **vascular bundles** in the stem. In the leaf, vascular tissues form a central large vascular bundle (midrib) from which smaller vascular bundles (veins) run through the leaf.

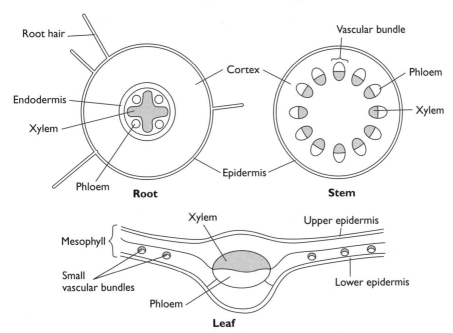

Figure 13 The distribution of vascular tissue (xylem and phloem) in cross sections of root, stem and leaf

Xylem

There are several different types of cell in xylem. The cells that transport most of the water and ions are called **xylem vessels**. In these cells, a secondary wall, impregnated with **lignin,** is formed inside the primary cellulose wall. Lignin is impermeable to water, so mature xylem vessels are dead.

There are different patterns of lignification in xylem vessels (Figure 14). In the first formed xylem, known as the **protoxylem**, produced in the growing regions behind the root and shoot tips, an **annular** or **spiral** pattern is produced. These patterns allow the vessels to elongate along with other tissues in the growth regions. In the xylem produced in the mature parts of the plant, known as the **metaxylem**, there is a greater deposition of lignin and a **reticulated** or **pitted** pattern is produced. Reticulated vessels are thickened by interconnecting bars of lignin; pitted vessels are uniformly thickened, except at pores seen as pits that allow rapid movement of water and ions out of the vessels to surrounding cells. The role of the lignin is to prevent the vessels from collapsing when under tension. As the vessels form their end-walls break down, so continuous tubes are formed. As a result, water movement through xylem vessels requires less pressure than through living cells where movement would be slowed down by cell contents.

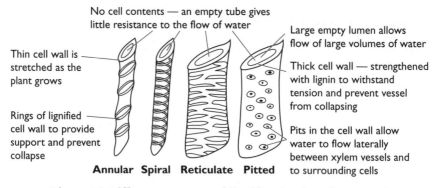

No cell contents — an empty tube gives little resistance to the flow of water

Large empty lumen allows flow of large volumes of water

Thin cell wall is stretched as the plant grows

Thick cell wall — strengthened with lignin to withstand tension and prevent vessel from collapsing

Rings of lignified cell wall to provide support and prevent collapse

Pits in the cell wall allow water to flow laterally between xylem vessels and to surrounding cells

Annular Spiral Reticulate Pitted

Figure 14 Different patterns of lignification in xylem vessels

Phloem

Phloem tissue consists of different cell types (Figure 15). The transporting cells are the **sieve tube elements**. These lie end-to-end to form a continuous stack — the sieve tube. The thin cellulose walls at the ends of the cells are perforated to form **sieve plates**, so making movement between sieve tube elements easier. **Companion cells** are closely associated with the sieve tube elements. The cytoplasm of the companion cell is linked via plasmodesmata to that of the sieve tube element. The smaller companion cells have a dense cytoplasm with a large number of mitochondria and have high levels of metabolic activity (see page 32). They do not transport materials but maintain the activity of the sieve tube elements. This is important since sieve tube elements lose their nuclei and many organelles as they mature.

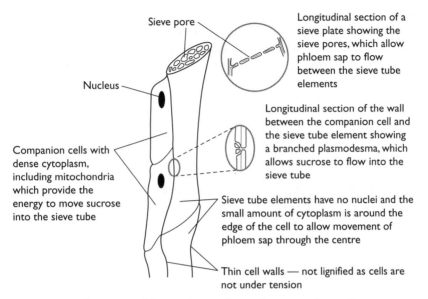

Sieve pore

Longitudinal section of a sieve plate showing the sieve pores, which allow phloem sap to flow between the sieve tube elements

Nucleus

Longitudinal section of the wall between the companion cell and the sieve tube element showing a branched plasmodesma, which allows sucrose to flow into the sieve tube

Companion cells with dense cytoplasm, including mitochondria which provide the energy to move sucrose into the sieve tube

Sieve tube elements have no nuclei and the small amount of cytoplasm is around the edge of the cell to allow movement of phloem sap through the centre

Thin cell walls — not lignified as cells are not under tension

Figure 15 Phloem sieve tube and companion cells

Transpiration and the movement of water and ions

Transpiration is the term used to define the process whereby water vapour is lost from plants (Figure 16). One of the main functions of the waxy cuticle is to waterproof the leaf's surface and so reduce the evaporation of water from the cells of the epidermis. While being an efficient barrier, the cuticle is not impermeable to the passage of water and some water is lost by **cuticular transpiration**. However, water evaporates readily from the cell walls of the mesophyll cells, and so water vapour accumulates in the air spaces. Stomata are closed at night, so little of this water vapour escapes. During the day, the stomata are open to allow the inward diffusion of carbon dioxide for photosynthesis. As a consequence, water vapour diffuses out. This **stomatal transpiration** is the combined effect of evaporation from the mesophyll surface and the diffusion of water vapour out of the open stomata. Stomatal transpiration accounts for 90% or more of all water loss in most plant species.

Internal factors that affect the rate of transpiration are:
- leaf surface area — increased surface area increases the rate of transpiration because more surface with stomata is exposed to the air
- stomatal density — the greater the density, the more stomata (per unit area) there are for water to diffuse out of the leaf
- cuticle thickness — the thicker the cuticle, the more effective it is at waterproofing and, therefore reducing cuticular transpiration.

A number of **external factors** influence transpiration rate:

- **Temperature** High temperatures increase the transpiration rate in two ways: greater heat energy increases the evaporation of water from the walls of mesophyll cells and increases the diffusion of water molecules out of the open stomata.
- **Air movements** Increased air movement outside the leaf increases the transpiration rate by blowing away the diffusion shells of water vapour that gather outside the open stomata. This increases the water potential gradient out of the leaf.
- **Humidity** Increased humidity decreases the rate of transpiration because when the atmosphere holds more water molecules the water potential gradient to the outside is reduced.
- **Light** Transpiration occurs more quickly in bright light than in the dark since stomata, the major route by which water vapour is lost, are open.
- **Soil water availability** If water is in short supply, plants are unable to replace the water lost in transpiration with water from the soil, and the stomata close, which reduces water loss by transpiration

As the water evaporates from the mesophyll cell walls it is replaced by water from the xylem vessels in the leaf. The flow of water through the mesophyll occurs by two main pathways:

- the **apoplast pathway** — water moves along the cellulose fibrils of the cell walls
- the **symplast pathway** — water diffuses from cell cytoplasm to cell cytoplasm through the plasmodesmata

Most of the water follows the apoplast pathway since this offers least resistance.

As water is drawn out of the xylem vessels, a **tension** (negative pressure) develops and a **water potential gradient** is established through the plant. Water moves through the plant because of this water potential gradient and because of **cohesive forces** between water molecules — the water column in the xylem vessel moves 'as one'. Further, there are **adhesive forces** between water and the cell walls, which tend to prevent the water column dropping down under gravity. This mechanism of water movement through the plant is known as the **cohesion-tension theory**. The pull caused by transpiration is great enough to cause the circumference of a tree to become smaller when transpiration is at its maximum (i.e. midday).

Water and ions are taken up at the root epidermis, where the root hair cells greatly increase the surface area. There are two routes:

- Water is absorbed into the root hair cells by osmosis; ions are absorbed by active transport (or, if the diffusion gradient is favourable, by facilitated diffusion). Subsequent movement across the cortex is along the symplast pathway.
- Water and ions are adsorbed onto the cellulose fibrils of the cell walls. Subsequent movement across the cortex is along the apoplast pathway.

Transport of water across the cells of the **cortex** of the root occurs in the same way as transport through the mesophyll of the leaf, i.e. mostly by the apoplast pathway and some by the symplast pathway. However, the cells of the **endodermis** have a **Casparian strip** that completely encircles each cell and is impermeable to water. This is a barrier to the apoplast pathway, so water can *only travel via the symplast* *pathway into xylem vessels.* This allows active control of the passage of water and dissolved ions. Ions are pumped by the cytoplasm of the endodermal cells into the xylem and water follows along the resultant water potential gradient, thus creating **root pressure**. Water also moves into the xylem due to the transpirational pull.

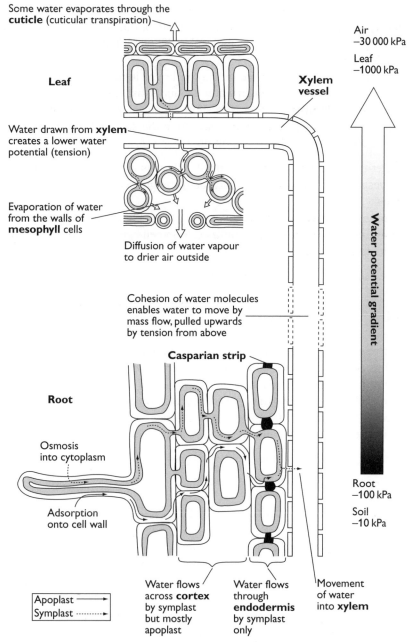

Some water evaporates through the **cuticle** (cuticular transpiration)

Air −30 000 kPa

Leaf −1000 kPa

Leaf

Xylem vessel

Water drawn from **xylem** creates a lower water potential (tension)

Evaporation of water from the walls of **mesophyll** cells

Diffusion of water vapour to drier air outside

Cohesion of water molecules enables water to move by mass flow, pulled upwards by tension from above

Water potential gradient

Casparian strip

Root

Osmosis into cytoplasm

Adsorption onto cell wall

Root −100 kPa

Soil −10 kPa

Apoplast ⟶
Symplast ┄┄▶

Water flows across **cortex** by symplast but mostly apoplast

Water flows through **endodermis** by symplast only

Movement of water into **xylem**

Figure 16 The movement of water through a plant

The movement of water out of a leaf, across a root and ultimately through the whole plant is shown in Figure 16, along with the water potential gradient from soil (less negative ψ) to air (more negative ψ).

Xerophytes

Xerophytes are plants that are adapted to living in dry habitats. Some of the features that enable plants to tolerate these dry habitats are summarised in Table 3.

Table 3 Summary of some xerophytic features

Features	Advantage of feature
Thick cuticle	Increases the efficiency of the waterproofing layer
Few stomata, leaf very small or with a reduced surface area-to-volume ratio	Reduced pathway for water loss
Stomata sunken into leaf; leaf covered by hairs; leaf rolled with stomata on the inside	Humid air builds up immediately outside the stomata reducing the diffusion gradient out of the leaf
Water storage cells in leaf or stem — succulence	Greater storage of water for use during periods of drought
Leaves adapted as spines	Spines prevent grazing and the exposure of plant tissue to evaporation
Extensive shallow root network or deep network of roots	Greater uptake of water when it does become available

Translocation of organic solutes

The movement of organic solutes within the phloem sap is known as **translocation**. Phloem sap contains primarily **sucrose** (the transport carbohydrate in plants), though amino acids and other solutes are also present. There are two main principles with respect to movement of phloem sap:

- Movement involves **energy expenditure**. There are several lines of evidence that mass flow is maintained by an active mechanism:
  - The rate of flow (1 m h⁻¹) is higher than can be accounted for by diffusion.
  - Companion cells have a particularly high density of mitochondria and are more metabolically active than other plant cells. ATP is used to pump sucrose into companion cells from where it enters the sieve tube element via plasmodesmata.
  - Metabolic poisons (e.g. potassium cyanide), which stop respiration, also stop translocation.
- Movement is **two-way** or, more precisely, movement occurs from '**source to sink**'. The source is the organ where sugar is produced in photosynthesis or by the breakdown of starch (e.g. leaves) and the sink is the organ that consumes or stores carbohydrate (developing buds, flowers, fruit, roots and root storage organs). So sugar can be moved up to a developing bud at the shoot tip or down to the roots. Furthermore, source and sink depend on the season: a potato tuber is a sink as it builds up stores of carbohydrate in the summer but is a source in the

spring when starch is broken down to supply the energy for the growth of shoots. Evidence for two-way flow comes from the use of radioactively labelled sucrose: following the supply of labelled sucrose to a mature leaf, radioactivity is detected in the shoot tip above and in the roots below.

Tip In an AS paper you may be given raw data and asked to construct a table. A **table-construction question** may ask you to 'organise the results into an appropriate table'. You will have decisions to make:

- Do the results need to be adjusted?
- What is the most appropriate caption to summarise the table contents?
- What is the best way to organise the table to make it easier to interpret the results?

Then, in constructing the table, you should ensure that there are explanatory column headings and that any units of measurement are included.

For example, in an experiment using a 'bubble' potometer, the water uptake of a leafy shoot under two sets of conditions is measured. Readings are taken, initially and thereafter at 2 minute intervals, of the position of the bubble on a millimetre scale. The data below is a record of the potometer readings.

Illuminated with a lamp: 3, 21, 39, 58, 77, 93

Covered with a black plastic bag: 9, 13, 17, 20, 24, 27

The initial reading, in each case, is best set to zero and the other readings adjusted accordingly (by subtracting the initial value from all the other values, i.e. subtract 3 from all values in the first set of conditions and 9 from all values in the second). An appropriate table, meeting the requirements in the mark scheme, is as follows:

	Potometer reading in two conditions/mm	
Time/minutes	Illuminated with a lamp	Covered with a black plastic bag
0	0	0
2	18	4
4	36	8
6	55	11
8	74	15
10	90	18

You may be asked to interpret and evaluate the results. In an interpretation, you would note that the lamp provides light ensuring that the stomata are open and so water molecules diffuse out of the leaf more readily (while heat from the lamp increases the rate of evaporation from the mesophyll surface), and that water uptake approximates to water loss in transpiration. When covered by a black plastic bag, the stomata close so that water is lost only via the cuticle (while humidity increases inside the bag and air currents are reduced, further reducing the water loss).

A common mistake made when interpreting potometer results is to discuss the influence of light on photosynthesis — photosynthesis uses only a small portion of the water taken up. Potometer results must be interpreted by discussing the influence of factors on transpiration.

An evaluation of the experiment should state that the results are not directly comparable since more that one factor is influenced by the treatments 'illuminated with a lamp' and 'covered with a black plastic bag'. Redesign the experiment.

Tip The topic of plant transport lends itself to the interpretation and/or drawing of photographs: a section through a root or the stele; a section through the stem or a vascular bundle; a section through any of a large variety of xerophytic leaves.

Practical work

Demonstration of a bubble 'potometer' and its use in measurement of the rate of water uptake:
- use to measure rate of water uptake
- use to investigate external factors that influence the rate of transpiration

Blood transport system in mammals

The mammalian circulatory system

In mammals, during one complete pathway around the body, the blood flows through the heart twice. One circuit going from the heart to the lungs and back is the **pulmonary circulation**. The other circuit from the heart to the rest of the body is the **systemic circulation**. Since there are two circuits, this is called a **double circulatory system**. Blood is pumped twice by the heart, from the *right ventricle to the lungs* and from the *left ventricle to the body*. These two chambers both have thick walls made of cardiac muscle, but the muscle of the left ventricle is thicker than the muscle of the right ventricle. Therefore, the blood pumped by the left ventricle is pushed harder and is under greater pressure than blood from the right ventricle. This two-pressure system has advantages:
- The low pressure in the pulmonary circulation pushes blood slowly to the lungs allowing more time for gas exchange.

- The high pressure in the systemic circulation ensures blood is pumped to all the other body organs and allows tissue fluid to form in each organ.

An artery branches off the systemic circulation to supply each of the body organs and a vein returns the blood to the heart. The main blood vessels of the thorax and abdomen are shown in Figure 17.

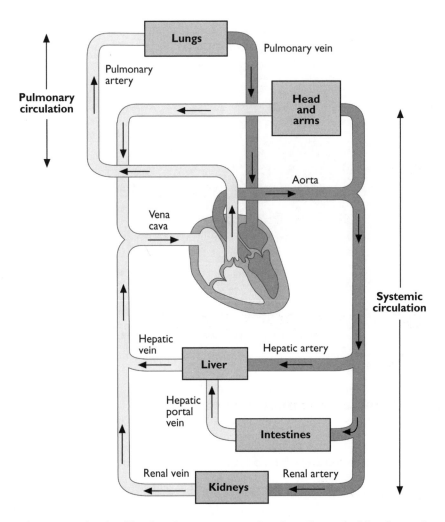

Figure 17 The double circulatory system showing the main blood vessels

The heart needs its own supply of blood — the coronary circulation — to provide cardiac muscle with oxygen and nutrients. The coronary arteries arise from the base of the aorta. Problems may occur in branches of the coronary arteries (see pp. 46–47).

The heart

The structure of the heart

The structure of the heart with its associated blood vessels is shown in Figure 18.

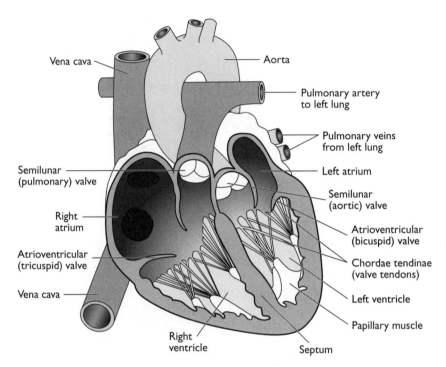

Figure 18 The structure of the heart

The cardiac cycle

There are three main stages during each beat of the heart:

- **atrial systole** — atria contract (ventricles are relaxed)
- **ventricular systole** — ventricles contract (atria are relaxed)
- **diastole** — both atria and ventricles are relaxed

These stages occur at the same time in the right and left sides of the heart.

The **pressure changes** in the left atrium, the left ventricle and in the aorta during one cardiac cycle are shown in Figure 19 on p. 38.

A graph of the changes in the right atrium, right ventricle and pulmonary artery shows all the same features, but the pressures in the right ventricle and pulmonary artery are lower.

The sequence of events during each phase of the cardiac cycle is shown in Table 4 on p. 39. The numbers of the stages in Table 4 are also shown in Figure 19.

As valves close, the flaps of tissue snap together making a sound. The first (and softer) heart sound is produced by closure of the atrioventricular valves; the second, sharper sound is produced by the clapping shut of the semilunar valves. These sounds are shown in the **phonocardiogram** in Figure 19 (see p. 38). The electrical activity through the heart is recorded as an **electrocardiogram** and is shown in Figure 19. Wave **P** shows the excitation of the atria, **QRS** indicates the excitation of the ventricles and **T** corresponds to diastole.

Coordination of the cardiac cycle

The sequence of atrial systole, ventricular systole and diastole occurs as a result of coordinated waves of excitation through the heart. The heartbeat is **myogenic** — contraction originates in the heart itself and does not depend on nervous stimulation. The heartbeat starts at the **sinoatrial node** (SA node), a small patch of tissue in the right atrium that acts as a **pacemaker**. Excitation from the SA node spreads rapidly over the atria causing them to contract together (atrial systole). Between the atria and the ventricles is a layer of **non-conductive tissue**, which prevents the spread of the wave of excitation passing directly from the atria to the ventricles. The only conducting route for the wave of excitation to the ventricles is via the **atrioventricular node** (AV node). Waves of excitation from the SA node reach the AV node and there is a short time delay before the waves of excitation pass down to the base of the ventricles. The AV node is connected to specialised muscle fibres called the **bundle of His**, which transports a wave of excitation down both sides of the septum to the base of the ventricles. Special fibres called **Purkinje fibres**, which branch upwards, conduct the waves of excitation to all parts of the ventricles causing them to contract from the bottom up. This ensures that blood is pushed up from the ventricles into the arteries. The two ventricles contract at exactly the same time (ventricular systole). Once cardiac muscle contracts it goes into a period of rest (diastole) before it can contract again. The waves of excitation through the heart are shown in Figure 20 on p. 40.

> **Tip** There are a number of things that you need to learn before you can have a good understanding of how the heart works. You need to:
> - learn the structure of the heart
> - understand the sequence of contractions
> - understand how this sequence is controlled by the electrical activity through the heart muscle
> - understand that as contraction occurs the pressure in a chamber increases, and that blood moves from an area of high to low pressure, if the valves allow it
> - understand that the action of valves depends on their structure, and that their opening and closure is governed by the pressure of blood — valves don't open or close independently, they are *forced* open or closed

You should be able to interpret data showing pressure changes in the heart chambers and major arteries, such as those in Figure 19. You need to work through this graph and recognise the events of the cardiac cycle. Note that blood flows from a region of high pressure to a region of lower pressure, unless prevented from doing so by the forced closure of a valve. When valve flaps snap together a heart sound is made. Figure 19 also shows a heartbeat of duration 0.8 seconds. This equates to a heart rate of 75 beats in 1 minute (60 × 0.8). A very fit person may have a heart rate of 60 beats per minute. Work out the duration of one beat.

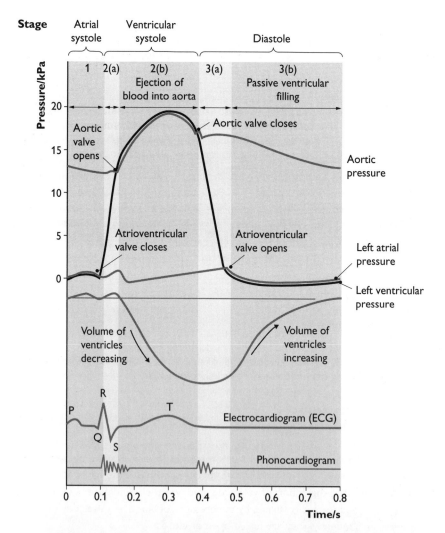

Figure 19 The pressure changes in left atrium, left ventricle and aorta during one cardiac cycle (with ventricular volume changes, electrocardiogram and phonocardiogram)

Table 4 The sequence of events during stages of the cardiac cycle

Stage	Description of events
(1) Atrial systole	Atria contract (ventricles are relaxed) pushing more blood into the ventricles. This is essentially topping up the ventricles (since blood has already entered the ventricles during diastole when the atrioventricular valves are open).
(2) Ventricular systole **(a)** **(b)**	Ventricles contract (atria are relaxed) causing the pressure of the blood inside the ventricles to become greater and forcing the atrioventricular valves shut. Two phases are evident (see Figure 19): **(a)** Ventricular pressure causes the atrioventricular valves to bulge into the atria increasing pressure there, while not being great enough to cause blood to exit the major arteries; the flaps of the atrioventricular valves are prevented from turning inside out by the chordae tendinae aided by contraction of the papillary muscles (see Figure 18) **(b)** Ventricular pressure increases to exceed that in the major arteries, pushing the semi-lunar valves open and causing the ejection of blood from the heart; blood is returned to the atria from the major veins and so atrial pressure gradually increases
(3) Diastole **(a)** **(b)**	Cardiac muscle throughout the atria and ventricles relaxes. Two phases are evident (see Figure 19): **(a)** Ventricular pressure drops to become lower than that in the main arteries so the semilunar valves are forced shut. Blood continues to be returned to the atria though it cannot enter the ventricles since the ventricular pressure is still greater than that in the atria, so atrioventricular valves remain closed. **(b)** Ventricular pressure drops to a point where it becomes lower than that in the atria. Therefore, the atrioventricular valves are forced open and blood enters the ventricles from the atria.

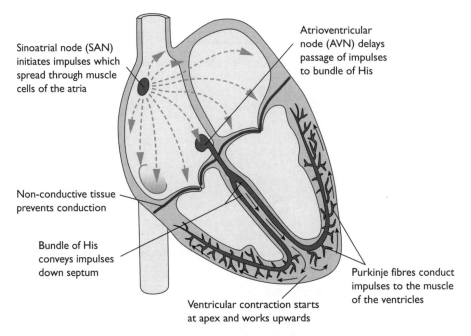

Sinoatrial node (SAN) initiates impulses which spread through muscle cells of the atria

Atrioventricular node (AVN) delays passage of impulses to bundle of His

Non-conductive tissue prevents conduction

Bundle of His conveys impulses down septum

Purkinje fibres conduct impulses to the muscle of the ventricles

Ventricular contraction starts at apex and works upwards

Figure 20 The waves of excitation through the heart

The blood vessels

Arteries are adapted for carrying blood under high pressure away from the heart, towards individual organs.

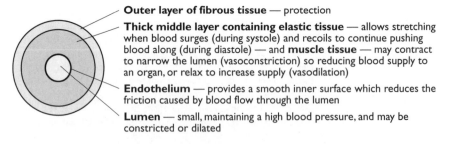

Outer layer of fibrous tissue — protection

Thick middle layer containing elastic tissue — allows stretching when blood surges (during systole) and recoils to continue pushing blood along (during diastole) — and **muscle tissue** — may contract to narrow the lumen (vasoconstriction) so reducing blood supply to an organ, or relax to increase supply (vasodilation)

Endothelium — provides a smooth inner surface which reduces the friction caused by blood flow through the lumen

Lumen — small, maintaining a high blood pressure, and may be constricted or dilated

Figure 21 The wall of an artery

Note that the muscle in arterial walls is *not* used to push the blood along.

Capillaries are adapted for the exchange of material between the blood and the tissue cells. The vast networks of capillaries slow the blood, giving time for the diffusion to occur.

The fluid that leaks out of the capillaries is called **tissue fluid**. It bathes all the surrounding cells. Polymorphs and monocytes, two types of white blood cell, can also leave the capillaries, squeezing between adjacent endothelial cells, at sites of infection.

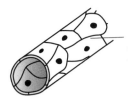

Squamous (pavement) endothelium — thin wall, permeable to water and solutes, so providing a short diffusion distance and facilitating the exchange of substances between the blood and tissue cells

Figure 22 A capillary

Veins are adapted for carrying blood under low pressure and returning it to the heart.

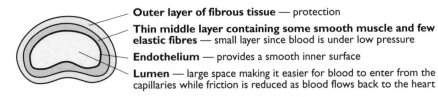

Outer layer of fibrous tissue — protection

Thin middle layer containing some smooth muscle and few elastic fibres — small layer since blood is under low pressure

Endothelium — provides a smooth inner surface

Lumen — large space making it easier for blood to enter from the capillaries while friction is reduced as blood flows back to the heart

Figure 23 The wall of a vein

Veins have semilunar valves to prevent backflow of blood. Blood is squeezed along when skeletal muscles contract.

Blood

Blood is a suspension of cells in a pale yellow liquid called **plasma**.

Blood cells

Erythrocytes (red blood cells) are much more numerous than white blood cells, of which there are different types — **polymorphs**, **monocytes** and **lymphocytes**. A drawing of a blood smear containing these cell types is shown in Figure 24.

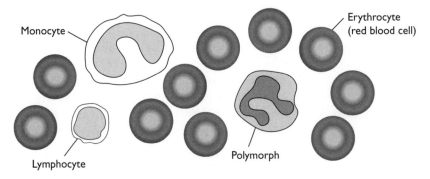

Monocyte

Erythrocyte (red blood cell)

Lymphocyte

Polymorph

Figure 24 A drawing of the different cell types in a blood smear

The structure and function of blood cells is summarised in Table 5.

Table 5 A summary of the structure and function of different blood cells

Cell type	Structure	Function
Erythrocytes (red blood cells)	Small (diameter 7–8 µm) cells lacking a nucleus and organelles, and with a biconcave disc shape. They are packed with haemoglobin.	The cells are adapted for the carriage of oxygen. The lack of a nucleus provides even more space for haemoglobin, an oxygen-carrying red pigment. The biconcave disc shape increases the surface area over which gas exchange can occur.
Polymorphs	Cells (diameter 10–12 µm) with a multilobed nucleus and granular cytoplasm. They are the most common white blood cell (70%).	They can squeeze between the endothelial cells of the capillaries at sites of infection where they engulf bacteria and other foreign bodies by phagocytosis.
Monocytes	Large cells (diameter 14–17 µm) with a kidney-shaped nucleus. The least common white blood cell (5%).	After moving out of the blood at sites of infection, they develop into macrophages which are long-lived phagocytic cells that engulf bacteria and foreign material.
Lymphocytes	Cells (diameter 7–8 µm) with a huge nucleus and little cytoplasm. Present in relatively large numbers (25% of white blood cells).	There are two types of lymphocyte: B lymphocytes are involved in antibody production as part of the immune response — antibody-mediated immunity; T lymphocytes are involved in destroying infected cells and foreign tissue — cell-mediated immunity.
Platelets	Essentially cell fragments too small to be readily visible using a light microscope.	Have an important role in initiating blood clotting and in plugging breaks in blood vessels.

Plasma

Plasma consists of 90% water and 10% of a variety of substances in solution and suspension: plasma proteins including prothrombin and fibrinogen (involved in clotting) and enzymes, hormones, glucose, amino acids, fats and fatty acids, urea (excretory product), vitamins and various ions (e.g. Ca^{2+} as a clotting factor).

Blood clotting

The clotting of blood seals cuts and wounds, preventing the entry of pathogens and stopping the loss of blood. When blood vessels are injured, a chain of reactions is initiated. In the final stages of the process, an inactive plasma protein, **prothrombin**, is converted into active **thrombin**. Thrombin is an enzyme that converts soluble **fibrinogen** into insoluble **fibrin**. The fibrin forms a mesh that traps red blood cells and forms a **clot**. A summary of the sequence of events in clotting is given in Figure 25.

If blood clots too readily then **thrombosis** may occur (an internal clot is called a **thrombus**). This is blood clotting in an unbroken blood vessel, which is dangerous and can lead to a stroke or heart attack. Conversely, if blood takes too long to clot **haemorrhage** may occur. In this case a great deal of blood may be lost from the blood vessels, which is also dangerous. The hereditary disorder **haemophilia** is a condition in which factor VIII is missing from the blood, as a result of which the blood cannot form clots without medical intervention.

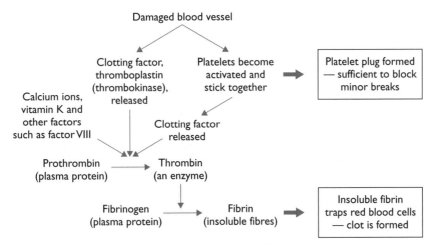

Figure 25 Blood clotting

Haemoglobin and the carriage of oxygen

Haemoglobin is a conjugated protein found in large quantities in red blood cells. Each molecule consists of four polypeptides: two α-chains and two β-chains. Each polypeptide has a **haem** group attached, which contains **iron** (Fe^{2+}). An oxygen molecule can associate with each haem to form oxyhaemoglobin.

$$Hb \quad + \quad 4O_2 \quad \rightleftharpoons \quad HbO_8$$
haemoglobin + oxygen $\rightleftharpoons$ oxyhaemoglobin

The equation shows that one molecule of oxyhaemoglobin can carry up to four molecules of oxygen. It also shows that the reaction is reversible: if oxygen is plentiful, oxyhaemoglobin is produced; if oxygen levels are low, oxyhaemoglobin breaks down (**dissociates**).

Blood contains a vast number of haemoglobin molecules —each mm³ of blood contains about 5 million red blood cells and each red blood cell contains about 280 million molecules of haemoglobin. The amount of oxygen carried by all this haemoglobin is measured as the degree to which the blood is **saturated** — 50% saturated means that, on average, each haemoglobin molecule is carrying two oxygen molecules. The amount of oxygen carried by the blood depends on the amount of oxygen available in its surroundings. This is measured as its **partial pressure** (abbreviated pO_2), which is the proportion of the total air pressure that is contributed by the oxygen in the mixture. It is measured in kilopascals (kPa). Figure 26 shows the **oxygen dissociation curve** of human haemoglobin. This indicates how the percentage saturation of blood with oxygen changes with the partial pressure of oxygen.

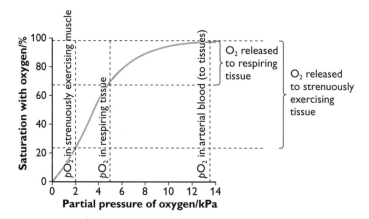

Figure 26 The oxygen dissociation curve of human haemoglobin

While the data for the graph are obtained experimentally (by bubbling air of different oxygen partial pressures though blood and measuring the degree to which the blood becomes saturated with oxygen) it provides evidence for what is happening in the body. The partial pressure of oxygen in the alveoli is relatively high (14 kPa). Therefore, since the blood entering the pulmonary circulation is deoxygenated, the haemoglobin **loads** with oxygen to become 98% saturated. In respiring tissues, oxygen is being used up and so the pO_2 is low (5 kPa). At this low pO_2, the blood can only be 70% saturated and, since blood arriving at the tissues is highly saturated, the haemoglobin **unloads** oxygen and so the tissues are supplied with oxygen. During strenuous exercise the pO_2 in muscles is reduced to 3 kPa, at which level the blood can only be 43% saturated, and so even more oxygen is unloaded. (You should use the graph in Figure 26 to go over these numbers.)

A significant feature of the graph in Figure 26 is that it is S-shaped (**sigmoid**). The reason for this relates to the way in which the four haem-containing polypeptides interact. When the first oxygen molecule combines with the first haem group, the shape of the haemoglobin molecule becomes distorted. This makes it easier for the other three oxygen molecules to bind with the other haem groups.

The amount of oxygen carried by haemoglobin depends not only on the partial pressure of oxygen but also on the **partial pressure of carbon dioxide (pCO_2)**. The effect of carbon dioxide on the oxygen dissociation curve of human haemoglobin is shown in Figure 27. This shows that at higher carbon dioxide partial pressures, the oxygen dissociation curve moves to the right. This is known as the **Bohr effect**.

The Bohr effect increases the efficiency of haemoglobin for oxygen transport. The partial pressure of carbon dioxide is high in tissues that are actively respiring (e.g. muscle). The higher the level of carbon dioxide, the lower the affinity of haemoglobin for oxygen, and so more oxygen is released to the tissues. This is advantageous because tissues with a high respiration rate require increased amounts of oxygen. In the alveoli the partial pressure of carbon dioxide is low (since it is breathed out), and so the affinity of haemoglobin for oxygen is increased and more oxygen can be loaded. This is advantageous because it allows the blood to become as fully saturated as possible.

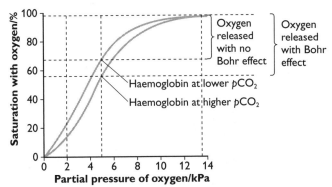

Figure 27 The effect of carbon dioxide on the oxygen dissociation curve of human haemoglobin

Higher temperature and lower pH also have a Bohr effect: exercising muscle generates more heat and, as the temperature in the muscle increases, more oxygen is released.

Myoglobin is not a blood pigment, but is found in 'red' muscle. It consists of one polypeptide with a single haem group, and does not, therefore, have a sigmoid dissociation curve. Myoglobin has a very high affinity for oxygen, as shown in Figure 28.

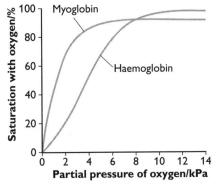

Figure 28 The oxygen dissociation curve of myoglobin compared with that of haemoglobin

Myoglobin only releases oxygen when the partial pressure of oxygen in the tissues becomes very low. It acts as an **oxygen store** within the muscle. If, as a result of strenuous exercise, the pO_2 becomes very low then myoglobin gives up its oxygen. This enables aerobic respiration to continue for longer and delays the onset of anaerobic respiration.

Oxyhaemoglobin dissociation curves for different species show considerable differences. For example, the oxygen dissociation curve for llama haemoglobin shows it to have a particularly high affinity for oxygen (see Figure 29).

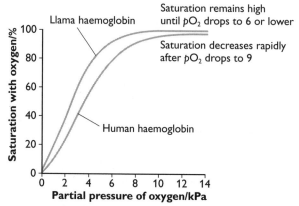

Figure 29 The oxygen dissociation curve of llama haemoglobin compared with that of human haemoglobin

The llama, a mammal of the camel family, is adapted to living at altitudes of about 5000 metres in the Andes mountains of South America. At high altitude, atmospheric pressure is low. The atmospheric pressure at sea level is 101.3 kPa so, with 21% oxygen present, the pO_2 in the air is about 21 kPa, though less at 14 kPa in the alveoli. At an altitude of 5500 metres the atmospheric pressure is halved. This means that pO_2 is reduced to 10.5 kPa and the alveolar pO_2 is only 7 kPa. The llama haemoglobin has a **high oxygen affinity** so that the blood can become fully saturated with oxygen at lower pO_2.

Human haemoglobin is not fully saturated at high altitude and an unacclimatised person would begin to show signs of lack of oxygen — breathlessness, headache, nausea and fatigue (mountain sickness). However, a person who moves to high altitude gradually, will, after 2–3 days, begin to become **acclimatised**. One of the most obvious changes is an increased production of red blood cells. Thus, there is a greater capacity for carrying oxygen to compensate for the lower levels of oxygen available. This response has been used by endurance athletes to increase their potential for oxygen delivery to the muscles when they compete subsequently at low altitude. However, the situation is not simple — for example, the quality of any training is reduced and must be less intense because of the lack of oxygen at high altitude.

Cardiovascular disease

Atherosclerosis

Atherosclerosis is a disease in which an artery wall thickens as a result of a build-up of fatty materials. Risk factors include smoking, inactivity, stress, salt intake, high blood cholesterol and alcohol consumption. The process involves the following sequence of events:

(1) The endothelium lining the artery becomes damaged. This damage can result from **high blood pressure**, which puts an extra strain on the layer of cells, or it might result from toxins from tobacco smoke in the bloodstream.

(2) Once the endothelium is breached there is an inflammatory response. Macrophages (derived from monocytes) leave the blood vessel and move into the artery wall. These cells accumulate chemicals from the blood, particularly **cholesterol**. A deposit builds up, which is called an **atheroma**.

(3) Calcium salts and fibrous tissue also build up at the site, resulting in a hard swelling called a **plaque** on the inner wall of the artery. The build-up of fibrous tissue means that the artery wall loses some of its elasticity — the artery is said to 'harden'. Atherosclerosis is often referred to as 'hardening of the arteries'.

(4) Plaques cause the artery to become narrow and, due to loss of elasticity, the artery is less compliant (i.e. unable to dilate or constrict). This makes it difficult for the heart to pump blood around the body and can lead to a rise in blood pressure. Raised blood pressure makes it more likely that further plaques will form.

If the arteries become very narrow or completely blocked then they cannot supply enough blood.

Coronary thrombosis

If a fatty plaque in an artery ruptures cholesterol is released, which leads to rapid clot formation. A clot that forms inside a damaged, but intact, blood vessel is called a **thrombus** and the condition is called **thrombosis**. If this happens in the **coronary arteries** it is called a **coronary thrombosis**. The heart muscle supplied by these arteries does not receive any oxygen. If the affected muscle cells are starved of oxygen for long they will be damaged permanently. This results in a heart attack or **myocardial infarction**. If a small branch of an artery is blocked only a small amount of muscle dies, causing a small heart attack; if a large artery is blocked, the whole heart may stop beating — a **cardiac arrest** (see Figure 30).

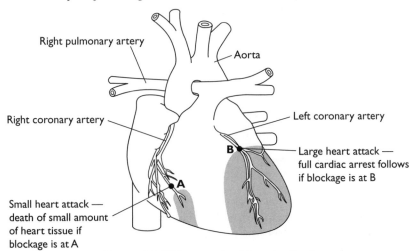

Figure 30 Different-sized heart attacks

Practical work

Examine prepared slides and/or photographs of blood vessels (in section) and mammalian heart (dissected and in section):
- distinguish between arteries, veins and capillaries
- identify heart chambers, atrioventricular valves, semilunar valves, chordae tendinae, papillary muscles, interventricular septum, major blood vessels (vena cavae, pulmonary arteries and aorta)

Examine stained blood films using the light microscope and/or photographs:
- identify erythrocytes, polymorphs, monocytes, lymphocytes and platelets.

Diversity of life

Classification

There is such a huge diversity of life on Earth that it makes sense to try to provide some degree of order by categorising the organisms — putting them into groups according to their similarities and differences. This grouping of organisms is known as **classification** and the study of biological classification is called **taxonomy**. Biological classification attempts to classify living organisms according to how closely related they are. It makes use of information from all areas of biology — for example cell structure, immunology, physiology, anatomy, behaviour, life cycles, ecology — and also from biochemistry. The basic unit of biological classification is the **species**.

The concept of the species

A species is defined as 'a group of organisms that is capable of interbreeding to produce viable and fertile offspring'.

This definition attempts to take into account some of the difficulties in describing 'what a species is':
- Robins in Ireland do not interbreed with robins in France, because they are geographically isolated, but they are *capable* of doing so — they are members of the same species.
- A horse and a donkey are capable of interbreeding, but their progeny, the mule, is *sterile* — they are different species.

Members of the same species have numerous features in common while still exhibiting some degree of variation. They have many genes in common, but variation is generated by the alleles that differ.

The number of species on Earth is estimated to be between 3 and 30 million. Only 2 million have so far been identified.

Classifying species

Species are classified into groups or categories of increasing size, i.e. **species**, **genus**, **family**, **order**, **class**, **phylum** and **kingdom**. Each of these groups is called a **taxon** (plural **taxa**). These are shown in Table 6.

Table 6

Taxon	Description
Genus (plural genera)	Group of related species
Family	Group of related genera
Order	Group of related families
Class	Group of related orders
Phylum (plural phyla)	Group of related classes
Kingdom	Group of related phyla

The Latin names of all taxa except that of species take initial upper case letters, while anglicised versions do not —for example the kingdom Animalia has members that are called animals.

Naming species

Species are named using the name of their genus and species. This is called the binomial system. By international convention, genus and species names are written in italics (underlined when handwritten); the genus name has an upper case initial letter and the species has a lower case initial letter, e.g. the European robin is named *Erithacus rubecula*. When a binomial has been used once in a passage it may be shortened subsequently (e.g. *E. rubecula*), providing that this does not cause confusion. If the species is not certain but the genus is known, it is possible to refer to an organism by the name of the genus followed by 'sp.' (plural 'spp.') — for example *Littorina* sp., a species of periwinkle.

The five kingdoms

A five kingdom classification system is presently recognised, though new discoveries lead to new ideas and this grouping may change in the future. These five kingdoms are:

- **Animalia**
- **Plantae**
- **Fungi**
- **Protoctista**
- **Prokaryotae**

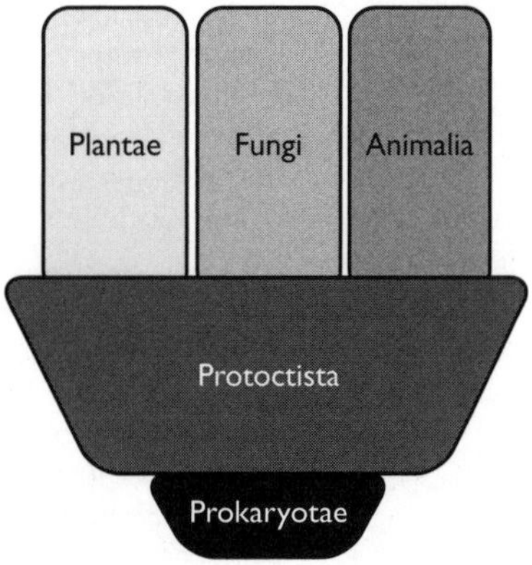

Figure 31 The five kingdom classification

Members of each kingdom have features in common, though some of these are also possessed by members of other kingdoms. For example, members of all the kingdoms except the Prokaryotae have eukaryotic cells. So, while having eukaryotic cells is a feature of the kingdom Plantae, it is *not* a distinguishing feature; members of the kingdom Plantae are distinguished by the possession of cellulose cell walls. Table 7 summarises the features of each of the five kingdoms, with distinguishing features shown in bold.

Table 7 The features of members of each of the five kingdoms

Phylum	Features (distinguishing features in bold)
Animalia	In all cases: multicellular; cells are eukaryotic; cells lack a cell wall. **All animals are heterotrophic** (consuming ready-made organic molecules)**; most ingest food into a digestive system.** They store carbohydrates as glycogen; and **usually store lipids as fats**. **Most are capable of locomotion.** Examples include flatworms, segmented worms, arthropods (e.g. insects) and chordates (fish, amphibians, reptiles, birds and mammals).
Plantae	In all cases: they are multicellular; cells are eukaryotic; **cells possess a cellulose cell wall.** All plants are photosynthetic, containing chlorophyll in chloroplasts, and are autotrophic (produce all their own food from inorganic material). They store carbohydrates as starch; and **lipids as oils**. Examples include mosses, ferns, conifers and flowering plants.
Fungi	Fungi are most often multicellular, though a few are unicellular (e.g. yeast). In all cases: cells are eukaryotic (most often organised into filaments, or hyphae, and frequently multinucleate and not divided into separate cells); and **cells have a cell wall, often made of chitin (*not* cellulose).** Fungi have a lysotrophic method of nutrition: they secrete enzymes to digest organic materials (usually dead, but some feed from living hosts) outside their cells and absorb the products of digestion. They store carbohydrates as glycogen and lipids as oils. Examples include yeast, moulds (e.g. *Mucor*) and toadstools.
Protoctista	This is a diverse group and membership is often by exclusion from all other groups. Some are unicellular, some are filamentous while others are multicellular though show limited differentiation. All protoctistans have eukaryotic cells. Some possess cell walls (cellulose or non-cellulose), chlorophyll (and can photosynthesise); others have no cell walls and are motile. They have a number of different methods of nutrition. Examples include seaweeds (e.g. *Fucus*) and *Amoeba*.
Prokaryotae	In all cases: **cells are microscopic and prokaryotic: they lack a nucleus and membrane-bound organelles; the DNA is naked and circular; ribosomes are smaller than those of eukaryotes; cell walls are made of peptidoglycans.** **Since nuclei are lacking cell division occurs by simple fission.** Different methods of nutrition are exhibited by the prokaryotes. Examples of this highly variable group include bacteria, blue-greens (Cyanobacteria) and hot spring bacteria (belonging to the Archaea group).

content guidance

Biological classification schemes are devised by taxonomists, based on the best available evidence at the time. New information, in 1990, giving greater weight to findings from molecular biology, has lead to the proposal of a completely new classification. As a result of this information the prokaryotes are now classified as two domains, the Archaea and the Bacteria. All other organisms (i.e. the eukaryotes) are classified within the domain Eukarya.

A phylogenetic classification

The most natural system of classification is one that reflects the ancestral or evolutionary relationships between groups. This system of classification is called **phylogenetic**. It puts the more closely related organisms together into the smaller groups, i.e. genus, then family, etc. Thus, the genus *Panthera* includes the closely related lion (*P. leo*) and tiger (*P. tigris*) but excludes the cheetah (*Acinonyx jubatus*) since it has unique features, although it possesses sufficient similarities to be included within the same family, Felidae.

Usually, the more features in common two species have, the more recently they have evolved from a common ancestor.

With the development of techniques for sequencing the nucleotides in DNA and RNA molecules and the amino acids in proteins, it has become possible to compare organisms at the most basic level — the gene. (Remember that the amino acid sequence in a protein is determined by the nucleotide sequence in a part of the DNA that is acting as a gene.) Closely related organisms are expected to possess a high degree of agreement in the molecular structure of their DNA, RNA and protein; the molecules of organisms related distantly usually show dissimilarity.

Protein analysis is used to compare the amino sequences of the same protein in different organisms. The more differences, the less closely related the species are presumed to be. Proteins that have been analysed include haemoglobin and cytochrome *c*, which is a protein of about 100 amino acids that is involved in aerobic respiration. Figure 32 shows the number of amino acid differences in the cytochrome *c* of six organisms and the phylogeny based on the cytochrome *c* data. While a classification cannot reliably be based on the analysis of a single protein, the results conform well to phylogenies based on other features.

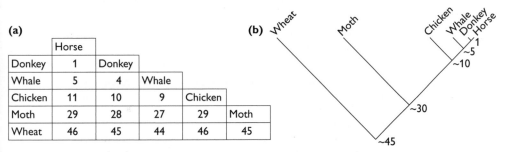

Figure 32 Cytochrome c analysis showing (a) the number of amino acid differences of six organisms, and (b) the resultant phylogeny

Identifying organisms using a key

Identification keys are useful for identifying unfamiliar organisms. They rely on a few, easy-to-see features. The most widely used key, the **dichotomous key**, separates organisms repeatedly into two groups until individual species have been described.

> **Tip** In the exam, you could be asked to construct a dichotomous key. A **dichotomous-key question** will probably present you with information about a number of different organisms. A useful starting point is to compile a comparative table of some obvious features shown by some of the organisms (if this has not been done for you in the question). For example, a table comparing four arthropods is shown below.

Feature	Housefly	Butterfly	Spider	Centipede
Segmentation	Yes	Yes	Yes	Yes
Legs	3 pairs	3 pairs	4 pairs	15 pairs
Wings	1 pair	2 pairs	None	None
Antennae	Short and stumpy	Long and club-shaped	None	Long and pointed

You will have decisions to make. Which features should you use to construct the key? In this example, 'segmentation' is of no use since it is common to all the organisms. In fact, only two features are sufficient to construct a dichotomous key. Then you must decide the order in which to use the features — this can be somewhat arbitrary. An example of a dichotomous key is shown below:

(1) Wings present → 2
   Wings absent → 3
(2) One pair of wings — housefly
   Two pairs of wings — butterfly
(3) Four pairs of legs -— spider
   15 pairs of legs — centipede

Guidelines to help you in the construction of keys include:
- Use clearly recognisable features (e.g. number of legs or number of leaflets in a leaf) and avoid those with a range of intermediates (e.g. degree of hairiness).
- Avoid colour and size because these may vary between male and female, adult and juvenile or from season to season.
- Avoid statements such as 'many' or 'few' — 'more than 12' or 'less than 5' are more precise.
- Avoid combining features (e.g. hairy leaves, white flower, stem with a square cross section). Try to keep to one aspect of a single feature (e.g. hairy leaves or leaves not hairy)

Biodiversity

Biodiversity is a contraction of 'biological diversity' and is used to describe the variety of life. It refers to the number *and* variety of organisms within a particular area, and has three components:

- species diversity
- ecosystem (or habitat) diversity
- genetic diversity

Species diversity

Species diversity refers to the number of different species and the numbers of individuals of each species within any one community. A number of objective measures have been created in order to measure species diversity.

Species richness is the number of different species present in an area. The more species present in a sample the 'richer' the area.

Species richness as a measure on its own takes no account of the number of individuals of each species present. **Simpson's index** (**D**) is a measure of diversity that takes into account both species richness and an evenness of abundance among the species present. In essence it measures the probability that two individuals selected randomly from an area will belong to the same species. The formula for calculating D is as follows:

$$D = \frac{\sum n_i(n_i - 1)}{N(N - 1)}$$

where $\sum$ = the sum of

n_i = the total number of organisms of each individual species

N = the total number of organisms of all species

The value of D ranges from 0 to 1. With this index, 0 represents infinite diversity and, 1, no diversity. That is, the bigger the value of D, the lower the diversity.

> **Tip** In the exam you could be asked to calculate Simpson's index. If so, the formula for Simpson's index, as presented above, will be provided within the question. (Note that some texts use derivations of the index above — as $1/D$ or $1 - D$. The index used here is the original equation as devised by Edward H. Simpson in 1949.) Worked examples of the use of Simpson's index to measure species diversity are provided on the biology microsite at www.ccea.org.uk.
>
> To calculate Simpson's index for a particular area, the area must be sampled and the number of individuals of each species noted. For example, the diversity of the ground flora in a woodland might be determined by sampling with random quadrats. The number of plant species in each quadrat and the number of individuals of each species should be noted. There is no necessity to be able to identify all the species, provided that they can be distinguished from each other. Percentage cover could be used to determine plant abundance but there must be consistency, i.e. either all by 'number of individuals' or all by 'percentage cover'.

Species biodiversity may be used to indicate the 'biological health' of a particular habitat. However, care should be used in interpreting biodiversity measures. Some habitats are stressful and so few organisms are adapted for life there, but those that are adapted may well be unique or, indeed, rare. Such habitats are important, even if there is little biodiversity. Nevertheless, if a habitat begins suddenly to lose its animal and plant types, ecologists become worried and search for causes, such as a pollution incident. Alternatively, an increase in the biodiversity of an area may mean that corrective measures have been effective.

Ecosystem diversity

This is the **diversity of ecosystems** or habitats within a particular area. A region possessing a wide variety of habitats is preferable and will include a much greater diversity of species than a region in which there are few different habitats. More specifically, countryside that has ponds, river, woodland, hedgerows, wet meadowland and set-aside grassland will be more species rich and more diverse than countryside with ploughed fields, drained land, without wet areas and devoid of woods and hedgerows.

Genetic diversity

This is the **genetic variability** of a species. It refers to the variety of alleles possessed by the individuals that make up any one species. Genetic diversity can be measured directly by genetic fingerprinting or indirectly by observing differences in the physical features of the organisms within the population (e.g. the different colour and banding patterns of the snail *Cepea nemoralis*). Genetic fingerprinting of individuals within cheetah populations has indicated very little genetic variability. Lack of genetic diversity is seen as problematic. It indicates that the species may not have sufficient adaptability and may not be able to survive an environmental hazard. The Irish potato blight of 1846, which resulted in the death of a million people and forced another million to emigrate, was the result of planting only two potato varieties, both of which were vulnerable to the potato blight fungus, *Phytophthora infestans*. The importance of genetic variation is discussed further on p. 58.

Adaptation of organisms

Organisms live in a **habitat**. They are part of an **ecosystem** within which they interact with both their **biotic** environment (the other living organisms) and with the **abiotic** environment (physical and chemical factors). Members of a single species form a **population** and, with other populations, make up a **community**.

Adaptations

Each organism has an **ecological niche** that describes its role within the ecosystem:
- what it feeds on or other nutrient needs
- what feeds on it
- competition with other organisms
- its temperature, water and other requirements

If it is to survive and reproduce, an organism must have adaptations for each of the above points. Adaptations may be behavioural, physiological or morphological. A **behavioural adaptation** is an aspect of the behaviour of an organism that helps it to survive and reproduce — for example, in attracting a mate. A **physiological** or **biochemical adaptation** is one in which there is appropriate functioning of the organism or its cellular processes —for example, the ability to respire anaerobically. A **morphological** or **anatomical adaptation** refers to any structure that enhances the survival of an organism —for example, the spines on a cactus that prevent grazing.

Adaptations of hydrophytes, plants specifically adapted for living in water, are shown in Figure 8 on p. 22.

Adaptations of xerophytes, plants adapted to living in dry habitats, are summarised in Table 3. This can be further expanded upon:
- Some xerophytes close their stomata when little water is available and some only open their stomata at night, when transpirational loss is reduced. These are behavioural adaptations.
- Some xerophytes possess cells that store water when it is readily available — the plants have succulent leaves and/or stems — for use in times of shortage. This is a physiological adaptation;
- Xerophytes have many morphological or anatomical adaptations such as sunken stomata that trap a layer of moist air next to the stomata, reducing the water potential gradient for water vapour to diffuse into the atmosphere.

Distribution of organisms

The distribution of organisms within their ecosystem is influenced, both separately and jointly, by biotic factors and abiotic factors, the latter being divided into climatic factors and edaphic (soil) factors.

Climatic factors include:
- **Temperature range** The Sun is the main source of heat for ecosystems. The temperature range within which life exists is relatively small. At low temperatures ice crystals may form within cells, causing physical disruption; at high temperatures enzymes are denatured. Fluctuation in environmental temperature is more extreme in terrestrial habitats than aquatic ones because the high heat capacity of water effectively buffers the temperature changes in aquatic habitats.
- **Availability of water** Water is essential to all life and its availability determines the distribution of terrestrial organisms. Many have adaptations to conserve water (e.g. the waxy cuticle of flowering plants and insects); some have less effective waterproofing and so are confined to moist or humid localities (e.g. mosses and woodlice). Even in aquatic habitats there may be problems due to the osmotic movement of water. In marine ecosystems, fish, for example mackerel and pollack, have adaptations to conserve water since there is a tendency for water to be withdrawn osmotically. Some fish (e.g. roach and perch) live exclusively in

freshwater and have adaptations to reduce osmotic gain. A few fish (e.g. salmon and eels) are capable of tolerating both extremes during their life cycles.

- **Light intensity** As the ultimate source of energy for ecosystems, light is a fundamental necessity. All plants require light for photosynthesis and most grow better the more light they receive. Some plants, however, are tolerant of low-light conditions, while some (e.g. bluebells) are adapted to grow on the woodland floor before the leaf canopy develops above. In aquatic ecosystems light is absorbed by the water molecules, so aquatic plants and algae can photosynthesise effectively only if they are near the surface. There is even less light in aquatic systems containing suspended particulates such as organic matter.
- **Light quality** Water does not only absorb light, it also influences which wavelengths can penetrate. Blue light penetrates water to a greater depth because red light is absorbed. Some marine algae, the red seaweeds, possess additional red pigments specifically to absorb at the blue end of the spectrum, and so red algae are adapted to live at greater depths than most other algae.
- **Day length** The longer the day length the more time a plant has for photosynthesis. The greater growth of plants in summer is more to do with the increased day length than it is to do with a rise in temperature. Day length also has a role in the flowering of plants.

Edaphic factors include:
- **pH values** The pH of a soil influences the availability of certain ions. Plants such as heathers grow best in acid soils, while spring gentian and cowslip prefer alkaline soils. Species that are tolerant to extremes of pH can become dominant in particular areas because competing species find it hard to survive in these extreme conditions. The dominance of heathers on upland moors is, in part, due to their ability to withstand very low soil pH. However, the optimum pH for the growth of most plants is close to neutral.
- **Availability of nutrients** There is a wide variety of ions required by plants. Some of these are needed in relatively large amounts and are called **macronutrients**: nitrate for the synthesis of amino acids etc., phosphate for the synthesis of nucleotides etc., calcium for the production of the middle lamella, sulphate for the synthesis of some amino acids and iron for the production of chlorophyll. Some ions are required in minute amounts and are called **micronutrients**. Different species make different demands on the ions in the soil and therefore plant distribution depends to some extent on the nutrient balance of a particular soil.
- **Water content** The water content of soils varies markedly. It depends on the soil type — clay soils tend to hold a lot of water, sandy soils are freely draining and hold little. A waterlogged soil creates anaerobic conditions. Plants able to tolerate these conditions include the rushes (*Juncus* spp.) and sedges (*Carex* spp.). They have air spaces within their root tissues that allow some diffusion of oxygen from the aerial parts to help supply the roots.
- **Aeration of soils** The space between soil particles is filed with air, from which the roots obtain their respiratory oxygen by diffusion. Soil air is also necessary to the aerobic microorganisms in the soil that decompose the humus.

Biotic factors include:

- **Competitors** Organisms compete with one another when they share a common resource, (e.g. food, water, light or ions) and that resource is in limited supply. They compete not only with members of other species — **interspecific competition** — but also with members of their own species — **intraspecific competition**. Where two species occupy the same ecological niche, the interspecific competition leads to the local extinction of one or the other — the **competitive exclusion principle**.
- **Predators** The distribution of a predator is reliant on the presence of its prey species. The population numbers of both prey and predator are interdependent —when prey numbers are low, predator numbers decline and, when predator numbers become high (due to abundant prey), then prey numbers drop.
- **Accumulation of waste** The growth of microorganisms is frequently self-limiting because the accumulation of waste products can be toxic —for example, in anaerobic conditions yeast populations produce ethanol.

An example of adaptation and distribution: rough and smooth periwinkles

The rough periwinkle (*Littorina saxatilis*) and the common (or edible) periwinkle (*Littorina littorea*) are marine snails found on rocky shores. Both graze algae on rocks and so competition between the two species might be considered to be high. However, the rough periwinkle is found on the upper shore, while the common periwinkle is prevalent in the middle shore. Conditions in the upper and middle shore areas are quite distinct. While the middle shore area is covered by tides for much of the time, the upper shore area is only covered for a few hours each day. Species living on the upper shore must survive the relatively dry conditions and the wide variation in temperatures that is possible. The result is that the two species, while being superficially very similar, possess a distinct series of adaptations to their quite different niches on the shore (Table 8).

Table 8 Adaptations of the rough and common periwinkle

Adaptation	Rough periwinkle	Common periwinkle
Behavioural adaptation	Sense of orientation that allows them to migrate back to their location on the upper shore if they are swept away by wave action	Able to migrate back to their location in the middle shore if they are swept away by tidal action
Physiological (or biochemical) adaptation	Adults retain fertilised eggs that hatch inside the body and are eventually released as small, shelled copies of the adults — developing eggs are protected from desiccation High temperature tolerance — able to survive the higher air temperatures in summer In extremes of desiccation they cement themselves onto rocks, respiring without oxygen for up to a week Excrete insoluble uric acid so that little water is lost	Egg capsules are shed into the water, eggs hatch into planktonic larvae (which drift in the tide) and eventually settle on the shore as small, shelled individuals — water is essential in the reproductive process Not tolerant of wide temperature variation — water temperature does not vary as much as air temperatures Excrete ammonia, which is soluble and must be excreted with lots of water

Adaptation	Rough periwinkle	Common periwinkle
Morphological (or anatomical) adaptation	Gills are modified to absorb oxygen from air — they are capable of surviving 1 month out of water	Gills are adapted for extracting oxygen from water — they can only breathe for a limited period out of water

Selection of organisms in populations

Environments are not uniform, i.e. factors are not constant throughout any particular habitat. For example, the soil-water content of a meadow may differ with some areas being simply moist and others very wet. (This provides different niches within the habitat and, therefore, areas for different species to survive and reproduce without being in competition.) Furthermore, the light intensity may differ across the meadow because different areas may have a different aspect — for example, some being on a south-facing slope, others facing north. However, populations (of the same species) vary and members of the same population may differ in the extent to which they are adapted to a particular set of conditions. For example, a particular plant species (such as the creeping buttercup, *Ranunculus repens*) may have some members that possess adaptations for very wet conditions while others have a preference for simply moist environments. This variation is genetic and the alleles causing the differences are passed on to any offspring.

There are processes in an organism's life cycle that generate this genetic variation:
- **Meiosis**, through the events of 'independent assortment' and 'crossing-over', produces haploid cells (e.g. gametes in animals), each of which is different from any other (see the AS 1 guide of this series).
- **Cross-fertilisation** brings a set of alleles from one parent into combination with a set of alleles from another parent.

Furthermore, new alleles may be introduced into a population by **mutation** (a spontaneous change in the coding DNA).

Within a relatively stable environment (in which conditions are not changing over time) most variations from the norm are selected against. The organisms most likely to reproduce successfully are those with characteristics close to the norm for the population — the **adaptive norm**. This is **stabilising selection**. It reduces the variation for the character since extreme variants are eliminated from the population. As a result, the constancy of the character is maintained.

If gradual changes occur in the environment over time the norm for the population may no longer be the most adaptive form. The variants towards one extreme may possess the fitter characteristics, so these are the organisms most likely to survive to reproduce and pass their genes on to the next generation. This is **directional selection**. There is a shift in the characteristic as different variants are selected for. A change in the genetic composition of the population results, i.e. there is **evolutionary change**. It allows the population to respond to environmental change and remain adaptive.

The two types of selection are described in Figure 33, for a continuous variable exhibiting a normal distribution

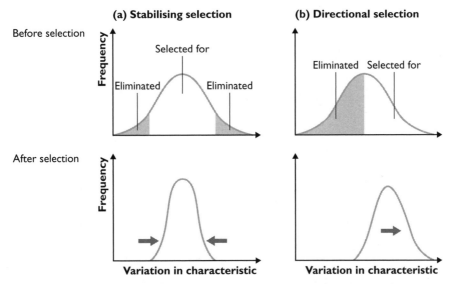

Figure 33 (a) Stabilising selection and (b) directional selection

Stabilising selection can be seen in the selection pressures on human birth weight. Measurements of the survival rate of babies in different weight classes have shown that mortality is highest among the groups weighing much less or much more than the average birth weight.

Directional selection was observed in peppered moths (*Biston betularia*) in Britain during the industrial revolution. Before the industrial revolution changed the environment, the light *'typica'* form was well adapted and the occasional dark *'carbonaria'* form (arising as a result of mutation) was selected against as it lacked camouflage on the lichen-covered trees. The environmental effects of the industrial revolution reversed these selective advantages, so that *'carbonaria'* became the favoured form with the result that its frequency increased in affected populations. Other examples of directional selection have been observed — for example, antibiotic resistance in bacteria, DDT resistance in mosquitoes and warfarin resistance in rats.

Practical work
Describe and carry out qualitative and quantitative techniques used to investigate the distribution and relative abundance of plants and animals in a habitat:
- sampling procedures to include:
  — random sampling
  — line transect
  — belt transect

- sampling devices to include quadrats, pitfall traps, sweep nets and pooters
- estimation of species abundance, density and percentage cover
- appreciation and, where possible, measurement of the biotic and abiotic factors that may be influencing the distribution of organisms

Human impact on biodiversity

Biodiversity describes the variety of living organisms on a local, national or global scale. **Global biodiversity** consists of approximately 2 million identified species of organisms in the **biosphere** (living world), and there are many more yet to be discovered. Over 50% of all known species are insects and most of these are species of beetle. When asked what could be inferred about the work of the Creator from a study of his works, the British biologist J. B. S. Haldane is reported to have replied 'an inordinate fondness for beetles'. Global biodiversity is influenced greatly by global warming (see A2 Unit 1). The emphasis in this chapter, however, is **local biodiversity** —the variety of life in Northern Ireland.

Threats to local biodiversity

Northern Ireland has lost over 50 species in the past century. Threats to biodiversity include habitat loss and fragmentation, including loss of native woodland, pesticide use and the establishment of invasive species.

Habitat loss and fragmentation

Habitats have been lost as a result of a number of activities relating to **intensive agricultural practices**, such as drainage schemes, removal of trees and hedges, and nutrient enrichment of soils. **Drainage schemes** were introduced to increase the area of agricultural land. As a result, many **wetland habitats** (e.g. rush pastures, raised bog, fens, reedbeds and wet woodlands) have been lost and these habitats are among the most important for wildlife — lapwing (*Vanellus vanellus*) numbers decreased by over 60% in the period 1987–99. The **removal of trees and hedgerows**, again to allow for more intensive agricultural practices, has also meant habitat loss for a variety of birds and other wildlife. The **nutrient enrichment of soils**, through the increased use of chemical fertilisers (containing the ions required by plants), has seen **species-rich hay meadows** decline by over 90% during the past 50 years. Many native grasses and herbs have become adapted to surviving in nutrient-poor soils (e.g. common bent grass, *Agrostis capillaris*, lady's bedstraw, *Gallium verum*, and the greater butterfly-orchid, *Platanthera chlorantha*). However, with the application of fertilisers, agriculturally preferred species of grass (e.g. perennial rye-grass, *Lolium perenne*) are able to grow much more quickly and out-compete the native plants. Anything that has a negative influence on plants has a similar effect on animal species. For example, one of the native plants in decline is devil's-bit scabious

(*Succisa pratensis*), which is the main food plant for the caterpillars of the marsh fritillary (*Euphydryas aurinia*), a butterfly similarly in decline.

Where areas of habitat are lost, fragmentation of small pockets of habitat can result. This means that some species become isolated as small separate populations within these pockets. This leads subsequently to inbreeding within the isolated populations and a loss of genetic diversity, leaving species less able to adapt to changes in their environment and ultimately increasing the risk of extinction. The previously extensive network of hedgerows acted as wildlife corridors for many species (e.g. small mammals, birds, reptiles, amphibians and insects), allowing dispersal and migration between habitats.

Loss of native woodland

The felling of trees for agricultural land, as fuel and as building material, occurred centuries ago. Northern Ireland is presently the least wooded region of Europe, with an approximate total tree cover of 6% (European average is 31%). Only 1% consists of mixed-species broadleaf semi-natural woodland (i.e. woodland which had some form of human modification in the past). The remaining 5% consists of introduced, exotic conifers (e.g. Sitka spruce, *Picea sitchensis* and lodgepole pine, *Pinus contorta*). Coniferous forests tend to have a thick floor of undecomposed needles, little light penetration and a soil that is acidic. As a result there are few plant species among the understory or ground layer. Since the trees have been introduced, few native animals are adapted to live there. Such conditions are not conducive to species biodiversity. There are, however, remnants of native forest — for example Banagher Glen Nature Reserve in County Londonderry, which contains mostly sessile oak (*Quercus petraea*), and Breen Wood Nature Reserve in County Antrim, which is dominated by sessile oak, but with some pedunculate oak (*Q. robur*), an understory of rowan (*Sorbus aucuparia*), hawthorn (*Crataegus monogyna*), holly (*Ilex aquifolium*) and hazel (*Corylus avellana*), and a ground layer with varieties of ferns, grasses and other flowering plants. These patches of woodland display enormous bio-diversity. You can get an appreciation of this diversity by visiting the website **www.virtualvisit-northernireland.com**, following the links 'see and do', 'nature and wildlife', then select 'Banagher Glen'.

Pesticides

Intensive agriculture was developed to maximise crop yield per unit area of land. It involved the creation of large open fields to make it easier to use mechanised machinery, and this also necessitated hedgerow removal. A single crop species is grown, which requires the addition of artificial fertilisers to maximise yields. As a **monoculture** (growing a single species), the crop plants are vulnerable to disease and pest problems, necessitating the application of chemical **pesticides**. Pesticides include **herbicides** (toxic to plants other than the crop), **fungicides** (toxic to fungi, e.g. cereal mildews) and **insecticides** (toxic to insects). The introduction of pesticides is associated with a huge loss in biodiversity — some farmland birds have declined by over 80%. There are several reasons:

- Pesticides may affect the soil ecosystem adversely by killing soil organisms. This will also adversely impact on other species in the food web.

- Pesticides may remove the natural enemies of pest species. This may lead to:
  — pesticide resistance if the pest species evolves
  — pest resurgence — the removal of a predator allows the pest to return in even greater numbers
  — secondary pest outbreak if a minor pest, previously kept at low numbers, multiplies rapidly in the absence of its competitor
- Herbicides reduce plant species diversity (with the removal of weed species in fields and in hedgerows if the spray drifts there). Therefore, the food available to many animal species in the food web is reduced, resulting in a lower diversity of animal species.

Invasive species

'Non-native invasive species' are species from other countries that establish themselves successfully in our habitats. They damage the 'native species' by out-competing them for habitat or food, preying on them, altering their habitat and by introducing disease and parasites. Invasive species usually do not have natural predators to control their numbers. For example, grey squirrels (*Sciurus carolinensis*), which are native to North America, threaten red squirrels (*S. vulgaris*) and woodland. They out-compete the red for food and are a known vector for the parapox virus that can be fatal to red squirrels. Grey squirrels cause economic loss to forest plantations by stripping the bark of young trees, which can lead to tree death.

Two species of flatworm, the New Zealand flatworm (*Artioposthia triangulata*) and the Australian flatworm (*Australoplana sanguinea var. alba*) have been introduced into Ireland. Both species predate earthworms and have significantly reduced earthworm numbers in some areas. This has had a hugely negative impact: earthworms are responsible for aerating the soil, decomposing plant material and are a food source for many animals, for example the song thrush, *Turdus philomelos*.

Rhododendrons are much admired garden plants, but *Rhododendron ponticum* is an invasive shrub that out-competes native species for resources, particularly light.

More information on invasive species can be found at www.invasivespeciesireland. com. After habitat destruction, invasive species are the second greatest threat to biodiversity worldwide.

Strategies used locally to encourage biodiversity

Woodland management

It is planned to increase tree cover in Northern Ireland from the present 6% and, in particular, to increase the cover by native trees. (A list of some native trees is provided in Table 9.) It is to be accomplished in a number of ways:

- encouraging farmers to plant native species, in hedges and field margins or in less fertile farmland
- restoring existing natural and semi-natural broadleaf woodlands
- establishing new plantations of native species — for example oak and ash, which will allow the growth of an understory of smaller trees, such as hawthorn and

rowan, which in turn will provide conditions for many species of ferns and flowering plants in the herb layer.

Native species provide a habitat and food for a large variety of insects and other invertebrates (see Table 9).

Table 9 The number of species of insects associated with Irish trees

Tree species	Tree type	Associated insect species
Oak (*Quercus* spp.)	Native, deciduous	284
Willow (*Salix* spp)	Native, deciduous	266
Birch (*Betula* spp.)	Native, deciduous	229
Hawthorn (*C. monogyna*)	Native, deciduous	149
Blackthorn (*Prunus spinosa*)	Native, deciduous	109
Scots pine (*Pinus sylvestris*)	Native, conifer	91
Ash (*Fraxinus excelsior*)	Native, deciduous	41
Spruce (*Picea* spp.)	Non-native, conifer	37
Sycamore (*Acer pseudoplatanus*)	Non-native, deciduous	15

Woodlands of native species support a more complex community and will have a massive effect on increasing biodiversity in Northern Ireland. This is not to say that conifers do not support biodiversity — for example, the red squirrel and the long-eared owl show a preference for coniferous woodland.

Hedge restoration and management

Farmers are encouraged to plant new hedges and replant hedges that have become gappy through neglect. A mixture of species must be planted (as a double row) and hedgerow trees must be included. The aim is to include a range of species and so provide a variety of niches for a large diversity of animal species.

Good hedgerow management through trimming also aims to promote biodiversity. The basic principles of good trimming are as follows:
- maintain a variety of hedge heights and widths to provide the optimum range of habitats (different bird species have particular preferences with respect to hedge height and width)
- trim in January or February to avoid destruction of birds' nests (March to August) and allow the berry crop to be used by wintering birds (September to December)
- trim on a 2- or 3-year rotation rather than annually, to boost the berry crop and insect populations
- avoid trimming all hedges in the same year (to allow hedge diversity)
- avoid cutting hedgerow trees (assuming they are native species)

Sustainable agricultural practices

There are trends now in agriculture to move away from monoculture, extensive applications of artificial fertilisers and ill-considered use of pesticides.

A traditional method was to grow different crops in one field over successive years — crop rotation. Crop rotation can improve soil fertility as different crops make different demands on soil nutrients. It can also help to avoid the development of diseases that are specific to only one crop. Polyculture is also being used. In polyculture crops are planted with a view to their mutual benefit. For example, a crop that attracts a natural predator of aphids helps a nearby crop that is infested with aphids. The need for pesticides is therefore eradicated.

The use of artificial fertilisers is now more judicious. Soils are tested for the level of different nutrients (i.e. ions) and fertiliser applied according to specific needs. There is also an increased use of organic fertiliser. This aids soil structure as well as releasing nutrients over time as it decomposes.

Broad-spectrum pesticides kill a wide variety of species and damage ecosystems:
- by killing the natural enemies of the pest
- by killing soil organisms
- by harming aquatic ecosystems through drift and run-off into waterways

Narrow-spectrum pesticides are toxic to a few species and so can be used more specifically to target the pest species.

It is important that a particular pesticide does not enter the food chain. For example, warfarin is a rodenticide used to kill rats that can damage grain stores. However, it has also been responsible for killing barn owls (*Tyto alba*). Now, new rodenticides are used that can be used in lower concentrations and are not harmful to barn owls.

Areas of farmland left undisturbed
Some farmers have **set-aside land**, which is never grazed or cultivated, to provide a habitat for a wide variety of species. They are also encouraged to leave an **ungrazed margin of grassland** at field perimeters. The aim is to provide additional habitat and food sources for a range of insects, birds and mammals. This strategy should contribute to the recovery of priority species such as linnet, yellowhammer, tree sparrow, barn owl and bats. It is also thought that these strips of land will encourage natural predators, for example some beetles, for pest control. For this reason, such margins are called **predator strips**. There is also support for farmers to plant native trees within the field margins. These strips extend the wildlife corridors along which many species can move between habitats.

Species-specific action
Each species has its own adaptations to its ecological niche. Therefore, there are specific considerations to be made with respect to individual species. There is a particular **action plan** for each threatened species, prepared nationally and at local council level. These detail the special considerations for improving numbers of the species in question. For example, some ground-nesting birds (e.g. the corncrake, *Crex crex*, now disappeared from Northern Ireland) have been affected detrimentally by the change from cutting grass towards the end of summer (to make hay) to making several cuts through the year (to make silage). Other species have been affected

adversely by the switch from spring planting of cereal crops to autumn planting, a practice that means ploughing takes place in the autumn and there is no winter stubble left with remnant grain for seed-eating birds. This has had a negative impact on yellowhammers (*Emberiza citrinella*) and tree sparrows (*Passer montanus*). The ground also has plant cover in the spring, which reduces nesting sites for skylarks (*Alauda arvensis*).

Habitat and species protection

Designating special areas for protection is an effective way of conserving the biodiversity of Northern Ireland. Different designations are shown in Table 10.

Table 10 Designations of sites for the protection of habitats and wildlife

Designation	Description
Areas of Special Scientific Interest (ASSI)	These have been identified by ecological surveys as containing valued plant and animal species, and are managed with the cooperation of the landowners. There are over 200 ASSIs in Northern Ireland (e.g. Drumlisaleen).
Special Areas of Conservation (SAC)	These are given special protection under the EU Habitats Directive to help conserve some of the most seriously threatened habitats and species across Europe. There are 54 SACs in Northern Ireland (e.g. Peatlands Park).
Special Protection Areas (SPA)	These are given special protection under the European Commission on the Protection of Wild Birds (the Birds Directive) as important areas for breeding, over-wintering and migrating birds. There are 15 SPAs in Northern Ireland (e.g. Rathlin Island).
Nature Reserves	These are areas of importance for plant and animal species and are directly managed for their conservation. There are almost 50 Nature Reserves in Northern Ireland (e.g. Banagher Glen Nature Reserve).
Marine Nature Reserve	The purpose of a Marine Nature Reserve is to conserve an exceptional range of marine organisms on the seashore and seabed. Northern Ireland has one Marine Nature Reserve, Strangford Lough.

More information on protected areas is available on the website www.ni-environment. gov.uk/protected_areas_home.htm.

Tip The section of the CCEA specification covering this topic is 2.3 (b), 'human impact on biodiversity', sub-sections 2.3.14 to 2.3.17. Each of these sub-sections begins with the verb 'appreciate'. This means that you should be familiar with the salient points of the topic, without having a detailed understanding of the area. For example, you might be asked to explain, in general terms, how biodiversity is adversely affected by 'annual cutting of hedgerows', 'use of pesticides', or 'increased use of artificial fertilisers', as described here. You cannot be expected to know the *specific actions* designed for the recovery of each threatened species — there are just too many species, each with its individual requirements — or to know the *details* of why a particular habitat is worthy of conservation. You may, however, be asked questions in these areas, but they will be based on your ability to apply understanding to information supplied, perhaps in the form of a comprehension passage, a table of data and/or a graph. As in all questions that ask you to apply your understanding, you must *read the information carefully* and *think* before answering. Nevertheless, your ability to work in unfamiliar areas may be improved by:

- undertaking a case study — for example, carrying out Internet searches on the action plans, local and national, for a particular species; there is information on this on the biology microsite at **www.ccea.org.uk**
- visiting a protected area near you and finding out why it is special — see the website **www.ni-environment.gov.uk/protected_areas home**.

Questions
&
Answers

This section consists of two exemplar papers constructed in the same way as your AS Unit 2 examination paper. There are questions that assess straightforward knowledge and understanding, some which require you to apply your understanding to novel situations and a few that assess your knowledge of practical techniques. There is a variety of question styles. Each paper has a total of 75 marks and you have 1 hour 30 minutes to attempt all the questions.

Following each question there are answers provided by two students — Candidate A and Candidate B. These are real responses. Candidate A has made mistakes that are often encountered by examiners and the overall performance might be expected to achieve a grade C or D. Candidate B has made fewer mistakes. The overall performance is consequently of grade A or B standard.

Examiner's comments

These are preceded by the icon e. They provide the correct answers and indicate where difficulties for the candidate occurred. Difficulties may include lack of detail, lack of clarity, misconceptions, irrelevance, poor reading of questions and mistaken meanings of examination terms. The comments suggest areas for improvement.

Using this section

You could simply read this section, but it is always better to be *active* in developing your examination technique. One way to achieve this would be to:
- try all the questions in Exemplar Paper 1, allowing yourself 1 hour 30 minutes, before looking at candidates' responses or the examiner's comments — remember to follow the suggestions in the introduction
- check your answers against the candidates' responses and the examiner's comments
- use the answers provided in the examiner's comments to mark your paper
- use the candidates' responses and the examiner's comments to check where your own performance might be improved

You should then repeat this for Exemplar Paper 2.

Section A

Question 1

If a blood vessel is broken (for example, following injury to the skin) thromboplastin is released and sets in motion a chain of reactions resulting in a blood clot.

(a) Complete the flow diagram below which summarises the clotting process.

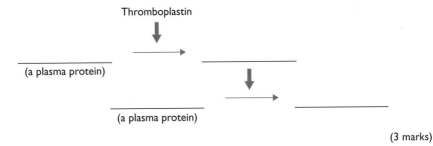

Thromboplastin

(a plasma protein)

(a plasma protein)

(3 marks)

Sometimes the clotting process may be initiated in an unbroken blood vessel. An internal clot, known as a thrombus, is formed. The condition is known as a thrombosis.

(b) Explain the possible consequences of a thrombus. (2 marks)

Total: 5 marks

Candidates' answers to Question 1

Candidate A
(a) Prothrombin ✓ → thrombin ✓ and fibrinogen ✓ → fibrin ✓

   This response is rewarded with full marks (i.e. 3 marks) for four correct answers.

(b) A thrombosis in the brain could cause a stroke ✓.

   This is correct, for 1 mark. To earn the second mark, some comment should have been made about either 'a thrombus causing a blockage' or the consequence of 'tissue not receiving oxygen'.

Candidate B
(a) Thromboplastin ✗ → thrombin ✓ and fibrinogen ✓ → fibrin ✓

   Thromboplastin is a clotting agent involved in the conversion of prothrombin to thrombin. Candidate B's three correct answers earn 2 marks.

(b) A thrombosis reduces blood flow and if it happened in a coronary artery the cardiac muscle would be deprived of oxygen ✓ and the person would suffer a cardiac infarction ✓.

This full answer with good use of the appropriate terminology scores both marks.

Overall, both candidates score 4 marks.

Question 2

The graphs below illustrate the effect of two different forms of selection (**A** and **B**) on the frequency of a characteristic within a population. In each case the upper graph represents the starting frequencies of the characteristic and the lower graph illustrates the outcome of the selection process.

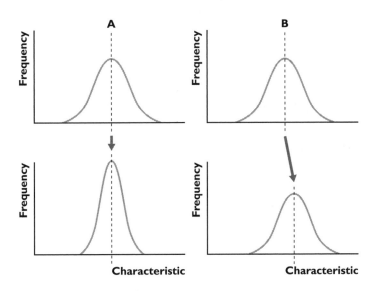

(a) Which of the two selection types, **A** or **B**, represents stabilising selection? Give a reason for your choice. (2 marks)

(b) Which of the two selection types, **A** or **B**, could result in a possible evolutionary change in the population? What is this type of selection known as? (2 marks)

Total: 4 marks

Candidates' answers to Question 2

Candidate A

(a) B ✗ — since the graphs are more or less the same shape ✗.

📝 It is graph A that represents stabilising selection. This is distinguished by a mode that remains the same, but where the extreme variants have been eliminated. The candidate fails to score.

(b) A ✗ — directional ✓

📝 It is graph B that shows a shift in the characteristic in the population. It is directional selection that leads to evolutionary change, for 1 mark.

Candidate B

(a) A is stabilising selection ✓ because a non-changing environment maintains the frequency of a desirable trait ✗.

> The choice of A is correct, for 1 mark. However, the modal forms are selected for and so their frequency increases; extreme forms are selected against and so their frequencies decrease.

(b) B is directional selection ✓ since a changing environment brings about a change in the frequency of the characteristic ✗.

> B is correct, for 1 mark. However the reasoning is not precise enough to earn the second mark. Directional selection occurs when there is a change in which *forms* (within the range for the characteristic) are favoured. It is not just a matter of a change in frequency, since any form of selection will do that. Of course at the level of the gene, directional selection causes a change in *allele* frequency.

> **Overall, Candidate A scores 1 mark and Candidate B scores 2.**

Question 3

Sand dunes may differ in their species composition. The diagram below is a vertical section through one particular sand dune system.

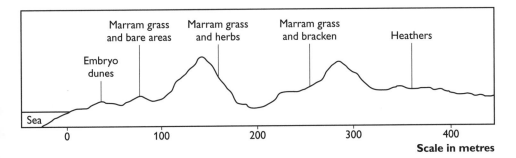

Describe how you would sample the area depicted in the diagram and use the results to illustrate quantitatively the zonation of the vegetation. (6 marks)

Total: 6 marks

Candidates' answers to Question 3

Candidate A

You would start at the embryo dunes and randomly sample ✓ the area using quadrats ✓. You would determine the percentage cover ✓ of each plant species. You would then move to the next area and continue ✓ until the landward area is reached. You would record the results.

> ✍ The candidate has provided the basis of a sampling procedure:
> - random sampling
> - sampling at sites transecting the dunes
> - use of quadrats
> - determination of percentage cover to measure abundance
>
> However, there is no reference to the identification of plant species — for example using a key — and no attempt to suggest how the zonation would be illustrated — for example by a series of kite diagrams (one for each species) with percentage cover (*y*-axis) plotted against position along the sand dunes. The question is not answered fully. The candidate needs to pay more attention when reading the question. He/she scores 4 marks.

Candidate B

Set a line of transect through the dunes ✓ and place quadrats ✓ along this at intervals ✓. Calculate the total percentage cover of all the plants in the quadrats ✗. Then put the results in a table and compare the different percentage covers of different zones within the sand dunes ✗.

There are only three correct points given here. It is not sufficient to determine the total percentage cover of all plants — the determination must be for each species. No attempt is made to suggest how the zonation would be illustrated – for example by a series of kite diagrams (one for each species) with percentage cover (*y*-axis) plotted against position along the sand dunes. Candidate B scores 3 marks.

Overall, Candidate A scores 4 marks and Candidate B scores 3.

Question 4

The diagram below shows a section through an alveolus and associated structures.

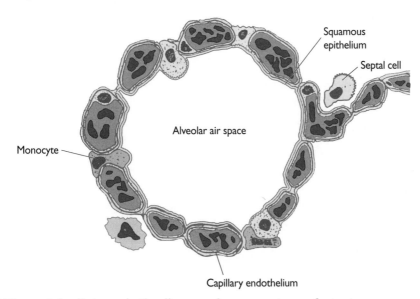

(a) **The septal cell shown in the diagram above secretes surfactant. What is the role of surfactant in the alveolus?** (1 mark)

(b) **State one adaptation, shown in the diagram, which facilitates gaseous exchange. Explain how this adaptation aids gaseous exchange.** (2 marks)

(c) **State one adaptation, *not* shown in the diagram, which is needed for efficient gaseous exchange. Explain how this adaptation aids gaseous exchange.** (2 marks)

(d) **State a role for the monocyte shown in the diagram.** (1 mark)

Total: 6 marks

Candidates' answers to Question 4

Candidate A

(a) To reduce the surface tension ✓

   ✎ This is sufficient to obtain the mark, though a fuller answer would note that a reduction in surface tension prevents the walls of the alveoli sticking together and collapsing during exhalation.

(b) Thin membranes ✗. This results in a greater rate of diffusion ✗.

The term 'membrane' can be confusing. It is not cell membranes that are thin but the cells of the alveolar walls — they form a squamous epithelium. The *rate* of diffusion is not altered — it is the distance over which the gases have to diffuse that is reduced. The important thing to remember here is to be precise. The candidate fails to score.

(c) There are millions of alveoli making up the lungs ✓ and these greatly increase the surface area over which gas exchange can take place ✓.

 The diagram does not show the vast number of alveoli, so this is correct, as is the explanation. Candidate A scores both marks.

(d) To digest any worn out organelles ✗

 It is lysosomes within cells that 'digest worn-out organelles'. Monocytes develop into macrophages that ingest any bacteria present. The candidate fails to score.

Candidate B

(a) Surfactant reduces the surface tension preventing the polar water molecules on the inner walls of the alveoli from sticking together ✓.

 This is correct, for 1 mark.

(b) The wall of the alveolus consists of a squamous epithelium in which the cells are thin ✓. This means that the diffusion distance is short ✓.

 This correct answer earns both marks.

(c) A layer of moisture lines each alveolus ✓. Gases must dissolve in water before they can diffuse through the cells ✓.

 This is correct, for 2 marks.

(d) The monocyte moves into the alveolus where it becomes a macrophage that engulfs any bacteria ✓.

 This is an excellent answer, which is well worth the mark.

 Overall, Candidate A scores 3 marks and Candidate B scores the full 6 marks.

Question 5

The yellowhammer, *Emberiza citrinella*, is a member of the bunting family. It feeds on grain, weed seed and grass seed in winter, though in summer adults and chicks feed mainly on invertebrates.

(a) The table below shows the classification of *Emberiza citrinella*. Complete the table by inserting the names of the three missing taxonomic groups.

Kingdom	Animalia
	Chordata
Class	Aves
	Passeriformes
Family	Emberizidae
	Emberiza
Species	E. citrinella

(3 marks)

(b) In Northern Ireland the yellowhammer has declined by 65% over the past 30 years, and is the subject of a Northern Ireland Species Action Plan. Recommendations to reverse this decline include the following:
- limit the use of herbicides
- reduce the use of insecticides
- restore areas of rough grass left undisturbed at field margins

Explain how each of these recommendations might aid the recovery of the yellowhammer population in Northern Ireland. (3 marks)

(c) As we have so little native woodland in Northern Ireland, hedges are an important substitute for woodland edge habitat. As such they host a wide range of nesting birds. Different species of bird have different preferences with respect to the structure of the hedge, as shown in the table below.

Species	Preferred height of nesting site in hedge	Preferred hedge width (if any)
Yellowhammer	Low (< 1 metre)	Wide (> 2 metres)
Bullfinch	Very high (> 4 metres)	Wide
Chaffinch	High	
Goldfinch	High	Wide
Greenfinch	High	Wide
Linnet	Low (< 1 metre)	

Use this information to explain why it would be preferable to trim farmland hedges on a 3-yearly rotational basis, rather than annually. **(2 marks)**

(d) Apart from providing nesting sites for birds, suggest two other benefits of hedges to the biodiversity of Northern Ireland. **(2 marks)**

Total: 10 marks

Candidates' answers to Question 5

Candidate A

(a) Phylum ✓, division ✗, genus ✓

> ✐ Phylum and genus are correct, for 2 marks. Division is another name (mostly used with the plant kingdom) for a phylum.

(b) Yellowhammers eat weed seeds in the winter and these may be covered in herbicides which could poison the yellowhammers. Insecticides may also be toxic ✓. Areas of rough grass left undisturbed would be free of herbicides and insecticides ✗.

> ✐ 1 mark has been allowed for the possible toxic effect of pesticides (herbicides and insecticides). The candidate fails to use the information provided in the question stem: that the yellowhammer relies on weed seed (which herbicides would remove) during the winter, relies on insects (killed by insecticides) as food during the summer, while areas of undisturbed rough grass would represent a source of both seeds and insects throughout the year.

(c) This would allow the hedges to grow upwards and outwards so providing birds with their preferred nesting sites ✓. Cutting annually may kill the hedges ✗.

> ✐ High, wide hedges are preferred by some species and so 1 mark is awarded. However, some bird species prefer low hedges. In essence, a 3-year rotational trimming regime should mean that, at any one time, some of the hedges (i.e. those recently cut) are low while some (those uncut) would be high and wide, i.e. there would be a variety of hedge types to suit different species.

(d) Hedges provide shelter for cattle ✗. They provide safe pathways for animals to move without attracting the attention of predators ✓.

> ✐ The first answer, while a correct statement, does not provide a benefit to 'biodiversity' as asked for in the question. The second is correct and earns 1 mark.

Candidate B

(a) Phylum ✓, order ✓, genus ✓

> ✐ The candidate scores all 3 marks.

(b) Herbicides kill weeds and so the yellowhammers would have no weed seeds to feed on in the winter ✓. Insecticides kill insects which yellowhammers feed on during summer ✓. Restoring areas of rough grassland provides a sanctuary for grass, weeds and insects which are important food sources for the yellowhammer ✓.

🖉 This excellent answer earns all 3 marks.

(c) Many species prefer high nesting sites and wide hedges and trimming the hedges every three years allows time for these to grow ✓. Short hedges do not suit four of the six species ✗.

🖉 High, wide hedges are preferred by some species and so 1 mark is awarded. However, some species prefer low hedges. A 3-year rotational trimming regime can be planned to produce a variety of hedge types to suit all species.

(d) Many hedgerow plants produce berries that are food for a variety of animals ✓. They represent wildlife corridors that allow animals to move safely between areas ✓.

🖉 This excellent response scores both marks.

🖉 **Overall, Candidate A scores 5 marks and Candidate B scores 9.**

Question 6

The photograph below is of part of a transverse section through the root of a buttercup (*Ranunculus*).

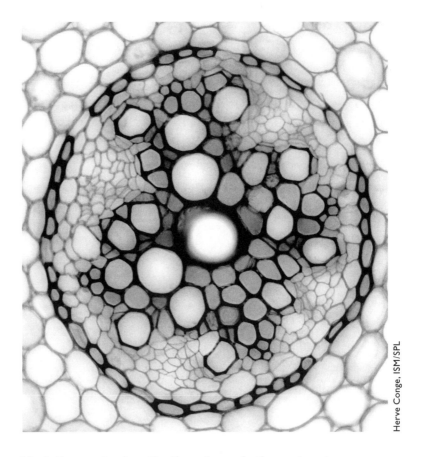

Herve Conge, ISM/SPL

Draw a block diagram to show the tissue layers in the root as shown in the photograph. Label the drawing to identify at least four structures. (9 marks)

Total: 9 marks

Candidates' answers to Question 6

Candidate A

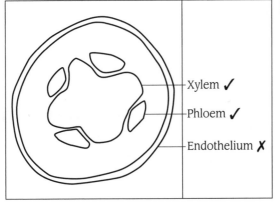

🖉 This is a block diagram showing tissue layers ✓ and has a degree of completeness in showing the tissues obvious in the photograph ✓.

It is also an attempt to represent the photograph ✓ though it lacks the proportionality of the features shown ✗ since the xylem region is too small within the stele.

The lines drawn are smooth and continuous, and not sketchy. ✓

The candidate gains 4 marks for drawing skills.

🖉 The labels for xylem and phloem are correct, for 2 marks. However, endothelium is not acceptable for endodermis. Other recognisable features include the stele (vascular cylinder) and the cortex.

Candidate B

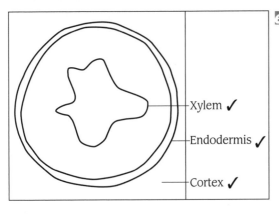

🖉 This is a block diagram showing tissue layers ✓ but only the xylem and endodermis are drawn and so it lacks completeness ✗.

It is an attempt to represent the photograph ✓, but it lacks the proportionality of the features shown ✗.

The lines drawn are smooth and continuous, not sketchy ✓.

The candidate gains 3 marks for drawing skills.

🖉 The candidate scores 3 marks for correctly labelling xylem, endodermis and cortex. The phloem is not drawn and so not recognised.

🖉 **Overall, both candidates score 6 marks.**

Question 7

Read the passage below and then use the information in the passage, and your own understanding, to answer the questions that follow.

The corncrake (*Crex crex*) is a small bird of the family Rallidae. It arrives in Ireland from mid-April, and the rasping call of the male, which is used to attract a mate, used to be one of the most familiar sounds of the summer. It no longer breeds in Northern Ireland. Secretive and difficult to see, corncrakes prefer to remain concealed in long grass, nettles and other tall vegetation. They feed on a range of invertebrates taken from plants or from the ground.

Nests are constructed on the ground from dead stems and leaves among the tall vegetation of hayfields and farm grassland. A first brood of ten eggs is produced in late May with peak hatching dates in mid-June. Females feed the chicks for the first few days and stay with them for about ten days. A second brood, again of ten eggs, is produced by the beginning of July with hatching taking place in late July.

A hundred years ago the population was measured in the tens of thousands. By the late 1960s the population had declined to less than 10000. Since then, all-Ireland censuses have been carried out by recording the number of 'singing' males. In 1988, only 916 singing males were heard; in 1993 the number of singing males had fallen to 174.

These declines correspond to the replacement of traditional farming systems by modern agricultural methods. Increasingly sophisticated machinery meant that grass could be cut earlier in the year and the harvest was completed more quickly than previously. Farmers also began to take several crops of grass per year. Earlier mowing destroys nests and kills chicks which are reluctant to leave the cover of tall grass, and may become trapped in an island of remaining grass at the centre of the field if mowing proceeds from the outside of the field inwards.

At present, corncrakes only breed in isolated pockets in the west of Ireland — for example on the islands off the coast of Donegal. By late September the surviving birds migrate to Africa where they overwinter before returning in the following spring.

(a) **Describe the ecological niche of the corncrake.** (2 marks)

(b) **The survival rate of chicks, prior to migrating to Africa, is 40% when reared in traditional hayfields, while the survival rate of adult birds from their migration in the autumn until their return in the following spring is 25%. Use this information to explain why corncrakes must raise two broods annually to be successful.** (2 marks)

(c) **Using the information in the passage about the numbers of calling males, calculate the percentage decline of corncrakes in Ireland in the 5-year period 1988 to 1993. (Show your working).** (2 marks)

(d) Suggest one reason for using 'the number of singing males' to estimate the numbers of corncrakes, and one reason why this method might not be valid. (2 marks)

(e) A number of schemes have been designed to develop favourable conditions for the improvement in corncrake numbers. Suggest two strategies that would be beneficial to corncrakes. (2 marks)

(f) With a small and fragmented population of corncrakes in Ireland there is a danger of inbreeding, which would result in a loss of genetic diversity. Explain why genetic diversity (variability) is important to populations. (2 marks)

Total: 12 marks

Candidates' answers to Question 7

Candidate A

(a) It nests in fields of tall vegetation ✓.

> ✏ The candidate describes an important aspect of the corncrake's habitat, but fails to expand on what it is doing there — feeding on invertebrates as a secondary consumer. Candidate A scores 1 mark.

(b) Because there is a chick survival rate of 40%, if only one brood was raised only four chicks would survive ✓. Then, if only 25% of the birds survive migration and return next spring this would only leave one bird. So two broods means that two birds will survive ✓, thus maintaining population numbers.

> ✏ This answer demonstrates a good understanding of the situation and explains it well (although reference might have been made to the need for two surviving birds to balance any adult losses). The candidate scores both marks.

(c) 174 (in 1993) ✗ ÷ 916 × 100 = 19% ✗

> ✏ The question asks for the percentage decline, i.e. the difference (916 − 174 = 742) divided by the starting number (i.e. 916) as a percentage (81%). The candidate fails to score.

(d) The birds are hard to see and so the males that sing can be heard and recorded easily ✓. Because there could be more females than males, not a 1:1 sex ratio, which would mean that there are more corncrakes than estimated ✓.

> ✏ These suggestions are well thought out and worthy of both marks.

(e) Farmers might only take one crop of grass a year. The grass could be cut later in the year when the chicks have left the nest ✓.

> ✏ These answers are not distinct, i.e. the second answer depends on the first. The second sentence scores 1 mark. See Candidate B's response for a fuller answer.

(f) Genetic differences allow the species to be more adaptable ✓.

> This is sufficient for 1 mark. For the second mark reference should be made to how genetic diversity might allow a species to survive an environmental change and be capable of evolving.

Candidate B

(a) The corncrake lives in long grass, nettles and other tall vegetation ✓. It feeds on a range of invertebrates taken from plants or the ground ✓, and it constructs nests on the ground.

> This excellent answer earns both marks.

(b) As most young corncrakes are wiped out during migration, a second brood must be raised so there are enough to maintain the population ✗.

> This answer has not used the relevant data to support the conclusion and the candidate fails to score. See Candidate A's answer for a fuller response.

(c) 916 (in 1993) less 174 (in 1993) = 742 ✓; 742 ÷ 916 × 100 = 81% ✓

> This answer is correct, for 2 marks.

(d) This is because males sing to attract a mate. Counting these shows the number of males capable of reproducing ✗.

> This is not really to the point. The reason for relying on the number of singing males is that the birds are so difficult to see. The problem with this is that it is not possible to know if all calling males are breeding (i.e. have been successful in attracting a mate) or if they attract and mate with one or more females. The candidate fails to score.

(e) Harvest the grass later in the year when both broods have hatched ✓ and promote set-aside grassland suitably managed ✓.

> Both of these are suitable suggestions, for 2 marks.

(f) If there were an outbreak of disease, genetic diversity would prevent all of the population from being wiped out ✓.

> This is sufficient for 1 mark. For a second mark, reference should be made to the possibility of *resistant forms* occurring within the variation shown.

> **Overall, both candidates score 7 marks.**

Question 8

The graph below shows pressure changes that take place in the left side of the heart during one complete cardiac cycle.

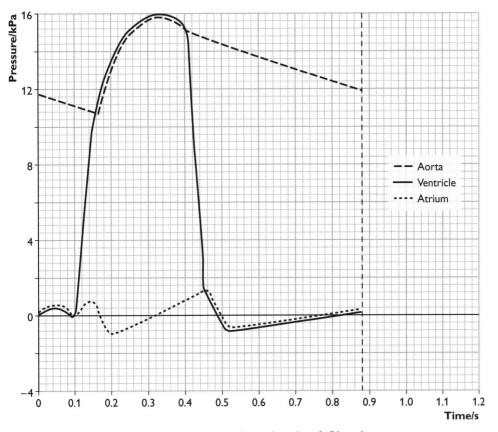

(a) At what time does the atrioventricular valve close? Give the reason for your answer. (2 marks)

(b) At what time does the aortic valve close? Give the reason for your answer. (2 marks)

(c) Explain the changes in the atrial pressure from 0.1 to 0.45 seconds. (2 marks)

(d) The pressure in the right ventricle reaches a maximum of 3.5 kPa. Give reasons for the difference between this pressure and the maximum pressure achieved in the left ventricle, as shown in the graph. (2 marks)

Total: 8 marks

Candidates' answers to Question 8

Candidate A

(a) 0.1 s ✓. This is when the atrium contracts, forcing blood into the ventricle and therefore increasing its pressure ✗.

> 🖉 The time is correct, for 1 mark. However, the reason given is irrelevant. The ventricle is contracting at this time, but even this is not the precise answer. At this time, the pressure in the ventricle increases to become greater than that in the atrium, so forcing the atrioventricular valve closed.

(b) 0.4 s ✓. This is when the aortic pressure is at its highest, so the valve closes to prevent backflow ✗.

> 🖉 Again, the time is correct, for 1 mark, but the reason given is not. Most particularly, a valve does *not* close to prevent backflow. In this case, it closes because the pressure in the ventricle drops below that in the aorta. As a result, backflow is prevented.

(c) Between 0.1 s and 0.12 s the atrium is contracting and pumping blood into the ventricle ✗. Thereafter, between 0.2 s and 0.45 s blood is returning to the atrium ✓.

> 🖉 The first part of this answer is incorrect. The second is correct, for 1 mark. The atrium contracted prior to 0.1 s. The peak in atrial pressure between 0.1 s and 0.14 s is due to the rise in ventricular pressure forcing the valve flaps to bulge into the atrial space (prior to blood forcing the aortic valve open).

(d) This is because the wall of the left ventricle contains more muscle ✓. This is required to produce a higher pressure to pump the blood around the whole body ✓.

> 🖉 Both parts are correct. An alternative answer would nave been to note that a low pulmonary pressure is required since slow blood flow increase the efficiency of gas exchange in the lungs, and also prevents fluid being forced into the alveoli. The candidate scores 2 marks.

Candidate B

(a) The atrioventricular valve closes at 0.1 s ✓, as contraction of the ventricle causes its pressure to increase above that in the atrium ✓.

> 🖉 Both the time and reasoning are correct. The candidate scores both marks.

(b) The aortic valve closes at 0.4 s ✓, as the ventricle relaxes and the ventricular pressure decreases and becomes less than that in the aorta ✓.

> 🖉 The candidate has a good understanding of the operation of the heart valves and scores both marks.

(c) There is a peak at 0.14 s as initially ventricular pressure increases though blood cannot exit the aorta and so the pressure pushes the flaps of the closed atrioventricular valve so that they dome into the atrium ✓. From 0.2 s, the pressure in the atrium increases gradually because the atrium is passively filling with blood from the pulmonary vein ✓.

🖉 Both statements are correct and indicate sound understanding. The candidate scores both marks.

(d) The muscular wall of the left ventricle is much thicker to pump blood into the aorta ✓.

🖉 This is correct for 1 mark, but no reference is made with respect to the *reason* for the higher pressure generated — see Candidate A's response and the examiner's comment.

🖉 **Overall, Candidate A scores 5 marks and Candidate B scores 7.**

Section A total for Candidate A: 34 marks out of 60

Section A total for Candidate B: 44 marks out of 60

Section B

Quality of written communication is awarded a maximum of 2 marks in this section.

(2 marks)

Question 9

Give an account of the processes involved in the movement of water through a plant.
(13 marks)

Total: 15 marks

Candidates' answers to Question 9

Candidate A

In the root, water travels from cell to cell via two routes: the symplast and the apoplast ✓. Apoplast is when the water travels through each cell ✗, whereas symplast is when water travels along the cell walls ✗. However, water cannot pass through the Casparian strip ✓ and so must enter the cell before passing into the xylem ✓. Water is then drawn upwards by negative tension, adhesion and cohesion. Negative tension is the force produced by water evaporating out of the leaves ✗. Adhesion is when the polar water molecules ✓ are attracted to the sides of the xylem ✓. Cohesion is when water molecules are attracted to each other ✓. Once the water travels up the stem it enters the leaves. The water is then passed from cell to cell and once it reaches the outer layer of cells next to the air space it evaporates ✓ and is released through the stomata.

> ✒ Seven appropriate points are provided, so Candidate A scores 7 marks. Some statements are simply wrong — for example, the symplast and apoplast routes have been confused. Other statements are not sufficiently precise: water does not evaporate out of the leaf; water vapour diffuses out following evaporation from the mesophyll surface. A number of relevant points (for example reference to the endodermis in the root) have been omitted. You should read Candidate B's answer for a fuller response.

> ✒ The candidate expresses ideas clearly, even though there are some factual errors, and the account is reasonably well sequenced. The candidate scores 2 marks for quality of written communication.

Candidate B

Water enters the root and moves through the cortex via the apoplast or symplast pathways ✓. Using the apoplast route, water moves through the cellulose cell walls ✓. This is the main route ✓ since there is less resistance to movement. Water moves through the cytoplasm of cells and from cell to cell via plasmodesmata ✓ using the symplast route ✓. Water can only move through the endodermis, and into the xylem, using the symplast route ✓.

Forces of adhesion between the water molecules and the lignified xylem vessels ✓ help the water to creep up the xylem ✓. The force of cohesion, due to the attraction between water molecules, maintains a continuous water column ✓. This water column moves up as water leaves the xylem to replace ✓ the water that has evaporated from the mesophyll surface ✓ and diffused out of the open stomata ✓. Throughout, water is moving along a water potential gradient ✓ as transpiration creates a particularly negative water potential in the leaf ✓.

> *The candidate receives a maximum 13 marks for the content.*

> *This is a well-structured account and ideas are expressed fluently. The links within the overall process are made clearly. The candidate scores 2 marks for quality of written communication.*

> **Overall, Candidate A scores 9 marks and Candidate B scores 15.**

Section B total for Candidate A: 9 marks out of 15

Section B total for Candidate B: 15 marks out of 15

Paper total for Candidate A: 43 marks out of 75

Paper total for Candidate B: 59 marks out of 75

Section A

Question 1

Use the dichotomous key below to identify the five kingdoms of living organisms. In each case, name the appropriate kingdom in the space provided.

(5 marks)

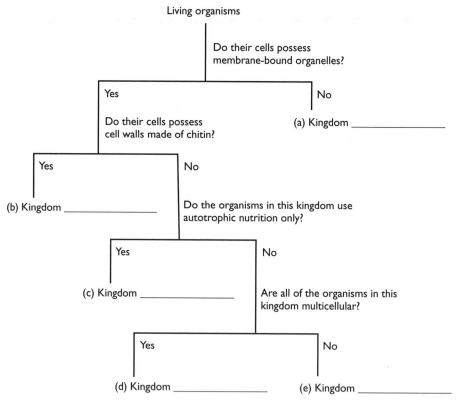

Total: 5 marks

Candidates' answers to Question 1

Candidate A

(a) Protoctista ✗
(b) Fungi ✓
(c) Plantae ✓
(d) Animalia ✓
(e) Prokaryotae ✗

🖉 The candidate has confused Protoctista with Prokaryotae. The other three kingdoms are correct, for 3 marks.

Candidate B

(a) Prokaryotae ✓
(b) Fungi ✓
(c) Plantae ✓
(d) Animalia ✓
(e) Protoctista ✓

🖉 Candidate B scores all 5 marks.

🖉 **Overall, Candidate A scores 3 marks and Candidate B scores 5.**

Question 2

The diagram below illustrates the structure of the leaf of a xerophyte.

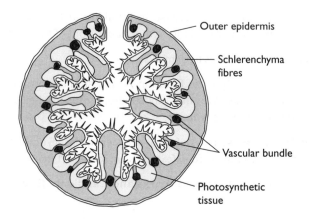

Outer epidermis

Schlerenchyma fibres

Vascular bundle

Photosynthetic tissue

(a) (i) **List two features, present in the diagram, which would operate to reduce transpiration.** (2 marks)

(ii) **Explain how one of the above features operates to reduce transpirational water loss.** (2 marks)

(b) **Identify two features of the leaf of a hydrophyte, and explain how each acts as an adaptation to environmental conditions.** (2 marks)

Total: 6 marks

Candidates' answers to Question 2

Candidate A

(a) (i) Leaf hairs ✓, rolled-up leaf ✓

    🖉 Candidate A has two correct features and scores 2 marks.

(ii) Leaf hairs trap humid air ✓ and reduce transpirational water loss ✗.

    🖉 The 'trapping' of humid air is correct for 1 mark: However, the consequence is not explained. The humid air outside the open stomata reduces the diffusion gradient for water vapour out of the leaf and so less is lost.

(b) Large prominent air spaces provide buoyancy ✓, and stomata on the upper surface allow gas exchange with the atmosphere ✓.

    🖉 The candidate gives two features that are correct for hydrophytes, the second only just sufficient (see Candidate B's answer for a better, fuller response) and scores 2 marks.

Candidate B

(a) (i) The leaf is folded or rolled inwards ✓, large air spaces ✗

    🖉 The first feature is correct, for 1 mark. However, the second is not — large air spaces are a hydrophytic adaptation.

(ii) The folded leaf causes an increase in humidity at the lower surface ✓ so reducing the diffusion gradient for water vapour out of the open stomata ✓.

    🖉 This excellent answer scores both marks.

(b) The leaf possesses hairs that aid flotation ✗, and stomata on the upper surface to allow for gas exchange with the atmosphere, a much richer source of gases, particularly oxygen, than the water ✓.

    🖉 The candidate shows some confusion between hydrophytic and xerophytic adaptations. However, the second feature is correct, for 1 mark.

    🖉 **Overall, Candidate A scores 5 marks and Candidate B scores 4.**

Question 3

(a) The diagram below shows an artery and vein in part of the pulmonary circulation.

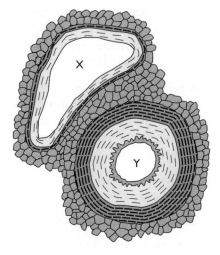

(i) Identify which of the two, **X** or **Y**, represents an artery. (1 mark)

(ii) Describe two features of an artery and explain how each is adaptive in the functioning of the artery. (2 marks)

(iii) Describe one feature of a vein and explain how it is adaptive in the functioning of the vein. (1 mark)

(b) Outline the sequence of events which result in the inhalation (inspiration) of air into the lungs. (4 marks)

Total: 8 marks

Candidates' answers to Question 3

Candidate A

(a) (i) Y ✓

> 🖉 This is correct, for 1 mark. The artery has a thicker wall and a smaller lumen than a vein.

(ii) Elastic fibres to withstand high pressure and allow recoil ✓; collagen fibres for strength ✗

> 🖉 The first adaptation is correct. However the reference to collagen is not a specific adaptation in the functioning of the artery. The artery has a thick layer of smooth muscle which, when it contracts, reduces the blood flow to the organ that it supplies. Candidate A scores 1 mark.

(iii) Possesses pocket valves ✗.

🖉 This is correct, but the candidate makes no attempt to explain why valves are adaptive — for example, that they ensure one-way flow towards the heart. The candidate fails to score.

(b) During inhalation the pressure in the lungs is lower than outside and so air rushes in ✓, causing the diaphragm to descend ✗. The external intercostal muscles contract, which pulls the rib cage upwards and outwards ✓. This increases the thoracic volume ✓, allowing the lungs to inflate.

🖉 There are three correct points here. However, there is confusion with respect to cause and effect within the inhalation process. The contraction of the diaphragm is a major cause of the increase in thoracic volume and in bringing about the inhalation of air — it does not result from inhalation. Candidate A scores 3 marks.

Candidate B

(a) (i) X ✗

🖉 This is incorrect, although from answers below this seems like a slip.

(ii) Elastic fibres to allow for expansion and recoil ✓; possesses a small lumen ✗

🖉 There is no attempt to explain why having a small lumen is an adaptive feature — it ensures that a high blood pressure is maintained. Candidate B scores only 1 mark.

(iii) They have a large lumen ✗.

🖉 There is no attempt to explain *why* a large lumen is an adaptation — for example, it provides little resistance to blood entering into and passing through the vein. The candidate gains only 1 of the 2 marks.

(b) The muscles of the diaphragm contract causing the diaphragm to move downwards ✓. The external intercostal muscles contract and the ribs are caused to move up and out ✓. This increases the volume of the thoracic cavity ✓. The pressure surrounding the lungs is therefore decreased ✓ with the result that air flows into the lungs along the pressure gradient.

🖉 This full answer scores all 4 marks.

🖉 **Overall, Candidate A scores 5 marks and Candidate B scores 6.**

Question 4

The graph below shows the frequency distribution of birth weights of infants in a London hospital between 1935 and 1946 (histogram) and the infant death rate in relation to birth weight (broken line). (Weight is shown in Imperial units, pounds; the unit is represented by lb.)

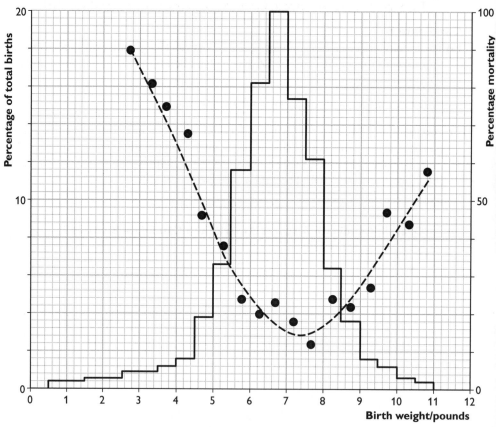

(a) Describe the trends shown in the graph. (3 marks)

(b) The data in the graph are often presented as an example of stabilising selection. Explain what is meant by stabilising selection, and its influence on the weight of infants. (4 marks)

Total: 7 marks

Candidates' answers to Question 4

Candidate A

(a) The closer the birth weight is to the mean, the fewer infant deaths there are ✓.

The candidate has not been penalised for using mean (rather than mode) and earns the mark. However, there are other relevant points that could have been made. These are provided in Candidate's B answer.

(b) Stabilising selection is apparent here since infants born close to the mean weight are favoured ✓. Those born further away from the mean weight have a reduced chance of survival ✓.

Both these points are correct. Candidate A scores 2 of the 4 marks available.

Candidate B

(a) The modal birth weight is between 6.5 and 7 lbs ✓. The greatest survival is among infants with these birth weights ✓. The highest death rate is among those infants with a birth weight at the extremes, i.e. below 5 lbs and over 9 lbs ✓.

This excellent description of trends scores all 3 marks.

(b) Stabilising selection occurs when individuals near the mode are favoured ✓ and extreme forms tend not to survive ✓. In this example, infants of 7 lbs have the greatest survival rate ✓, so the constancy of the characteristic is maintained and the modal birth weight will remain at approximately 7 lbs ✓.

This is an excellent answer, which scores the full 4 marks.

Overall, Candidate A scores 3 marks and Candidate B scores 7.

Question 5

(a) Explain what happens to the partial pressure of oxygen and carbon dioxide within the blood capillaries in the lungs and in actively respiring tissues, such as working muscle. (2 marks)

(b) The graph below shows oxygen dissociation curves for haemoglobin at two different partial pressures of carbon dioxide (pCO_2), and for myoglobin. The partial pressures of oxygen (pO_2) in the lungs, in muscle during moderate exercise, and in muscle during strenuous exercise, are indicated.

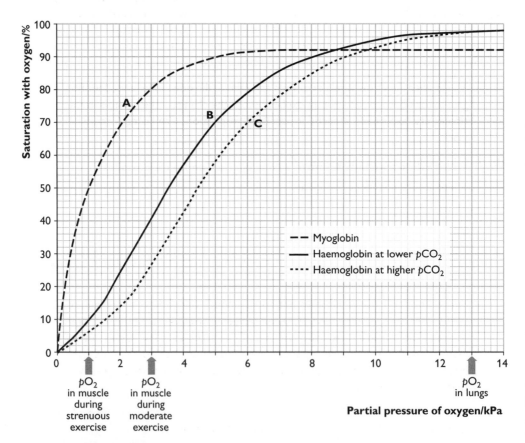

(i) Which of the curves, B or C, represents the dissociation curve for haemoglobin in the lungs and in respiring tissues? (1 mark)

(ii) What term is used to describe the effect of higher pCO_2 on the oxygen curve for haemoglobin? (1 mark)

(iii) Using your understanding of oxygen dissociation curves, and the information provided in the graph, explain why haemoglobin unloads oxygen to actively respiring tissues, such as working muscle. (3 marks)

(iv) Using your understanding, and the information provided in the graph, explain the role of myoglobin in a muscle. (2 marks)

Total: **9 marks**

Candidates' answers to Question 5

Candidate A

(a) In the lungs, the partial pressure of oxygen increases, while the partial pressure of carbon dioxide decreases ✗. In the tissues, the partial pressure of oxygen decreases since oxygen is being used in respiration, which also produces carbon dioxide and so the partial pressure of carbon dioxide increases ✓.

🗒 The answer about the levels of oxygen and carbon dioxide in the lungs does not gain a mark because it is not an explanation of the changes. The changes in the tissues are explained correctly in terms of respiration. Candidate A scores 1 mark.

(b) (i) B in the lungs, C in the tissues ✓

🗒 These are both correct, for 1 mark.

(ii) Lower affinity ✗

🗒 This is incorrect — it is called the Bohr effect.

(iii) In the tissues, oxygen is used up and in such circumstances haemoglobin unloads its oxygen ✓.

🗒 This is fine for 1 mark, but does not go far enough since there is no mention of the effects of an increase in pCO_2 on oxygen unloading. Part (a) is relevant here and the candidate has answered that pCO_2 increases in the tissues due to respiration. Remember that questions are often structured so that one part builds on another. This part-question is worth 3 marks and a fuller answer than that provided is necessary.

(iv) Myoglobin has a high affinity for oxygen and can take up oxygen more easily ✗. It enables aerobic respiration to continue for longer ✓.

🗒 The first part of the answer is wrong since it is describing what happens in the lungs, not, as required, in muscle. The second part is correct, for 1 mark.

Candidate B

(a) In the lungs, pO_2 increases while pCO_2 decreases, since oxygen is diffusing out of, and carbon dioxide into, the alveoli according to the diffusion gradients ✓. In the tissues, pO_2 decreases as oxygen is being used in respiration in the tissues cells, while pCO_2 increases because carbon dioxide is produced ✓.

🖉 The candidate notes the changes and provides explanations as required by the question. He/she scores both marks.

(b) (i) B in the lungs, C in the tissues ✓

🖉 These are both correct, for 1 mark.

(ii) Bohr effect ✓

🖉 This is correct, for 1 mark.

(iii) In respiring tissues, oxygen is used up and pO_2 is reduced to 3 kPa and, during strenuous exercise, this can become as low as 1 kPa ✓. At lower pO_2 haemoglobin cannot hold on to its oxygen and so it is unloaded ✓. Respiration produces carbon dioxide and so pCO_2 increases with the effect that haemoglobin releases even more oxygen ✓.

🖉 This is an excellent answer that contains more than three correct points. Note that the candidate makes appropriate use of the data in the graph. Candidate B scores all 3 marks.

(iv) Myoglobin binds with oxygen and only releases it at very low pO_2 ✓. It acts as an oxygen store ✓ in muscle and enables aerobic respiration to take place for longer.

🖉 This excellent answer scores both marks.

🖉 **Overall, Candidate A scores 4 marks and Candidate B scores 9.**

Question 6

The diagram below shows a simple bubble potometer. The potometer is used to measure the rate of water uptake by a cut, leafy shoot.

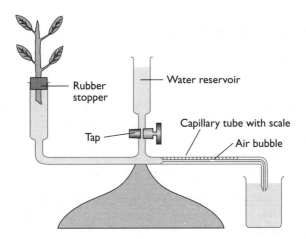

(a) (i) Describe how the potometer, as shown in the diagram, would have been set up. (3 marks)

   (ii) Describe how the potometer would be used to make several measurements of the rate of water uptake. (2 marks)

(b) State one major assumption that is made when using this apparatus to measure the transpirational loss of water. (1 mark)

(c) In order to investigate the effect of light on the rate of transpiration the following two experiments were devised:
   • Experiment 1: a bench lamp was placed near to the shoot
   • Experiment 2: a black plastic bag was placed over the shoot

   The following results were obtained:
   • Experiment 1: at time 0 the bubble was at 5 mm, and readings taken thereafter, at 2 minute intervals, were 20, 35, 50, 65, 79 and 93 mm
   • Experiment 2: the initial reading was 17 mm and subsequent readings of 21, 25, 29 and 33 mm were taken at intervals of 3 minutes

   (i) Organise the results into an appropriate table. Your table should have a caption, suitable column headings, show units and include all the data in such a way as to make interpretation easy. (5 marks)

   (ii) Explain, as far as possible, the results obtained. (2 marks)

(iii) Evaluate the validity of the experimental design in investigating the effect of light on the rate of transpiration. (2 marks)

Total: 15 marks

Candidates' answers to Question 6

Candidate A

(a) (i) The potometer is set up under water ✓. The end of the shoot is also immersed and the last centimetre of stalk is cut under water ✓. Vaseline is smeared round the base of the shoot and at the stopper in order to ensure that it is airtight ✓.

🖉 Three correct points are given, for 3 marks.

(ii) Measure the distance travelled by the air bubble ✓ and reset the air bubble by opening the water reservoir ✓.

🖉 The candidate scores both marks.

(b) The transpirational loss of water is equal to water uptake ✓.

🖉 This is correct, for 1 mark.

(c) (i)

Time/ minutes	Experiment 1 Distance on scale/mm	Experiment 2 Distance on scale/mm
0	5	17
1		
2	20	
3		21
4	35	
5		
6	50	25
7		
8	65	
9		29
10	79	
11		
12	93	33

🖉 There is no caption describing the contents of the table ✗.

The column headings are not explanatory — the experiments could be about any topic ✗, although the correct units of measurement are shown ✓.

Nevertheless, the table does have a logical construction ✓ and all the data are included at the correct time intervals ✓.

🖉 The candidate scores 3 marks for table construction.

(ii) With the bench lamp, there was more water uptake due to photosynthesis occurring ✗. When a black plastic bag was covering the plant, only respiration was occurring so little water was taken up ✗.

🖉 In part (b), the candidate has established that this apparatus is essentially measuring transpiration, but ignores this here. References to photosynthesis and respiration do not explain the results. The candidate fails to score.

(iii) As rate of water uptake equals rate of transpiration, the experiment is valid ✗. However, the minute intervals were not the same for both, making comparison more difficult ✗.

🖉 The first statement relates to the validity of using the apparatus (as answered in part (b)), and not to experimental design. Taking measurements at different times may mean that greater care is required when tabulating the results or plotting a graph, but it does not affect the validity. This is about there being more than one different variable in each experiment:
- the lamp provides heat as well as light
- being covered by a black plastic bag means that, as well as light not being available, there are no air currents and so humidity increases

🖉 The candidate fails to score.

Candidate B

(a) (i) The shoot should be inserted into the potometer while it is fully immersed in water ✓. Vaseline may be used to create an airtight seal ✓.

🖉 There are two correct points, for 2 marks. Other points could have been made — for example immersing the whole apparatus to remove any air.

(ii) The distance moved by the air bubble is measured over a fixed time (e.g. 10 minutes) ✓. The experiment is then repeated by opening the reservoir and re-setting the air bubble ✓.

🖉 The candidate scores both marks.

(b) That water lost by transpiration is replaced equally by water taken up ✓.

🖉 This is correct, for 1 mark.

(c) (i)

	Time/minutes	
	Experiment 1	**Experiment 2**
Distance moved by air bubble/ mm	0...2...4...6 ...8...10...12	0...3...6...9...12
	5...20...35...50 ...65...79...93	17...21...25 ...29...33

🖉 There is no caption, describing the contents of the table ✗.

The column headings are not explanatory — the experiments could be about any topic ✗, although the correct units of measurement are shown ✓.

The table lacks logical construction ✗, although the data may be supposed to be at the correct time intervals ✓.

Table construction is poor and the candidate scores only 2 marks. This is a skill to be practised (see the 'tip' on p. 33 in the section on 'Transport and transpiration in flowering plants.')

(ii) Placing the bench lamp near the shoot means that the stomata are open and water vapour may escape ✓. Covering the shoot with a bag increases the humidity inside and reduces the diffusion gradient so less water is lost be transpiration ✓.

Both statements are correct. The candidate scores 2 marks.

(iii) The lamp should have been placed a fixed distance from the plant. All other variables (temperature ✓, humidity) should have remained constant.

This is correct for experiment 1, but the candidate offers no answer for experiment 2 and, therefore, scores only 1 mark.

Overall, Candidate A scores 9 marks and Candidate B scores 10.

Question 7

Most of the plantation forest in Ireland is conifer, mainly Sitka spruce (*Picea sitchensis*). A Sitka spruce forest was surveyed for nesting bird species. The numbers of birds of each of four species found are shown in the table below.

Nesting bird species	Number of individuals
Goldcrest (*Regulus regulus*)	8
Siskin (*Carduelis spinus*)	8
Sparrowhawk (*Accipiter nisus*)	2
Treecreeper (*Certhia familiaris*)	4

(a) Calculate the value for Simpson's diversity index (*D*) for nesting bird species in the survey of the Sitka spruce forest. The formula for *D* is:

$$D = \frac{\sum n_i(n_i - 1)}{N(N-1)}$$

where n_i = the total number of organisms of each individual species

N = the total number of organisms of all species

(Show your working.)   (3 marks)

In a mixed forest of Sitka spruce and ash (*Fraxinus excelsior*) a survey of nesting bird species indicated that there were 18 species present giving a Simpson's Index (*D*) of 0.06.

(b) Use this information to explain the term 'species richness'.   (1 mark)

(c) Explain the difference between the values of Simpson's index (*D*) for the Sitka spruce forest and the mixed Sitka spruce/ash forest. Both forests were of comparable size.   (2 marks)

(d) What do the results suggest about the strategy for planting future forests?   (2 marks)

(e) Sitka spruce is an 'introduced species'. State two problems associated with introduced species.   (2 marks)

                                                                 Total: 10 marks

Candidates' answers to Question 7

Candidate A
(a) $D = [8(8 - 1) + 2(2 - 1) + 8(8 - 1) + 4(4 - 1)]$ ✓ $\div 20(20 - 1)$ ✗ $= 126 \div 380 = 0.33$ ✓

   The numerator part of the calculation is correct. The candidate has made a mistake in totalling the numbers column, which should be 22, not 20. However, the subsequent part is not penalised, since the arithmetic operation is correct and this can be seen clearly as the working out is shown. The mistake was a slip that would have been found if the candidate had taken time to check the arithmetic. Candidate A scores 2 marks.

(b) There are a lot of different species ✓.

🖉 This is correct, for 1 mark.

(c) The mixed woodland has a bigger value for Simpson's index (*D*) ✗. It is not possible to comment on the amount of diversity ✗.

🖉 This is not correct. *D* for the mixed woodland is smaller (given in the stem of the question as 0.06). A smaller value for *D* indicates greater diversity. The candidate fails to score.

(d) Forests should be planted according to their commercial value ✗.

🖉 This might be a correct point of view, but it is not appropriate here. The question asks for a consideration of the results, i.e. the increased diversity of mixed woodland. The candidate fails to score.

(e) Out-competing native species ✓.

🖉 This is correct for 1 mark. Other correct responses include:
- lack of natural predators and so a lack of control
- lack of association with native species and so a threat to biodiversity
- elimination of native species by, for example, predation
- hybridisation with native species

Candidate B

(a) $D = [8(8 - 1) + 2(2 - 1) + 8(8 - 1) + 4(4 - 1)]$ ✓ $\div 22(22 - 1)$ ✓ $= 0.27$ ✓

🖉 This is correct, for all 3 marks.

(b) This is the number of species in one area ✓.

🖉 This is correct, for 1 mark.

(c) The Sitka spruce forest has a lower species diversity than the mixed forest ✓. A greater number of plant species is present, so there is a greater variety of possible nesting sites and food available for different bird species ✓.

🖉 This full answer scores both marks.

(d) It is preferable to have forests planted with a mixture of tree species ✓. This would increase the number of available niches for a variety of animal species, especially native species ✓.

🖉 This is an excellent answer. Candidate B scores 2 marks.

(e) Few organisms can feed on them so the variety of organisms in the area is reduced ✓. New diseases are introduced so native species may be wiped out ✓.

🖉 There are two correct points here, for 2 marks.

🖉 **Overall, Candidate A scores 4 marks and Candidate B scores 10.**

Section A total for Candidate A: 33 marks out of 60

Section A total for Candidate B: 51 marks out of 60

Section B

Quality of written communication is awarded a maximum of 2 marks in this section.

(2 marks)

Question 8

Give an account of the phases of diastole, atrial systole and ventricular systole during the cardiac cycle. The account should make reference to each of the following:
- **the waves of excitation**
- **the pressure changes**
- **the opening and closure of valves** (13 marks)

Total: 15 marks

Candidates' answers to Question 8

Candidate A

Diastole: This phase of the cardiac cycle is the resting phase. On the right side of the heart blood flows into the right atrium and on the left side it flows into the left atrium ✗.

> There is nothing incorrect here, it just lacks the detail necessary to earn a mark. For example, where is the blood flowing from (the major veins), what happens to the pressure in the atria, what effect has this on the atrioventricular valves? The candidate fails to score.

Atrial systole: During atrial systole the right atrium fills with blood from the venae cavae and the left atrium from the pulmonary veins ✗. The atria contract and increase pressure in the atria forcing blood into the ventricles ✓, since when the pressure in the atria is greater than in the ventricles the atrioventricular valves open ✓. Atrial systole begins with electrical activity being produced at the sinoatrial node ✓.

> The atria fill with blood during diastole, not during atrial systole, and so the first statement is wrong. The second statement is correct. The opening of the atrioventricular valves actually occurs during diastole, but the concept is correct and so the third statement is awarded a mark. Then, there is a correct reference to the SA node initiating the wave of excitation. Candidate A scores 3 marks.

Ventricular systole: During ventricular systole, the tricuspid and bicuspid valves are forced shut ✓ and prevent backflow into the atria. The semilunar valves are forced open ✓. This allows oxygenated blood from the left ventricle to enter the aorta, from where it will supply the organs of the body; blood from the right ventricle enters the pulmonary artery and is taken to the lungs ✓. The semilunar valves slam shut and the pocket-like valves prevent backflow to the ventricles ✗.

📝 The candidate notes correctly that the atrivntricular valves are forced shut, and that, at a later time, the semilunar valves are forced open. However, there is no reference to the increasing pressure in the ventricles, that the atrioventricular valves are only forced shut when the pressure is greater than that in the atria and that the semilunar valves are forced open when the pressure in the ventricles is greater than in the major arteries. Semilunar valves close during diastole, when the pressure in the ventricles becomes lower than in the major arteries. Overall, while there is good understanding of the structure of the heart, there is a lack of understanding of the pressure changes and the electrical activity that coordinates the contractions. The candidate scores 3 marks here, giving a total of 6 marks for the biological content of this answer.

📝 The ideas are not well sequenced, so the candidate scores only 1 mark for quality of written communication.

Candidate B

Diastole: During diastole the heart is relaxed ✗.

📝 This is not sufficient for a mark here. However, some events of diastole are discussed appropriately later in the answer.

Atrial systole: During atrial systole an electrical impulse is generated within the sinoatrial node ✓. This wave of excitation propagates over the walls of the atria and causes contraction ✓. This results in the remaining blood within the atria being pushed into the ventricles ✓. Blood flows down the pressure gradient and the atrioventricular valves are pushed open.

📝 There are three correct points here, for 3 marks, although some detail is missing. The AV valves are already open during atrial systole, since the atrial pressure increases when blood is returned to the heart.

Ventricular systole: The wave of excitation reaches the atrioventricular node ✓. The electrical impulses then spread downwards through the bundle of His ✓ and the Purkinje fibres, stimulating contraction of both ventricles from the apex upwards and pushing blood towards the arteries ✓. As the pressure in the ventricles increases, the atrioventricular valves are slammed shut ✓. As the pressure increases further it becomes greater than in the arteries and the semilunar valves are forced open ✓.

📝 This is a reasonable account of ventricular systole though, again, some detail is missing — for example, there is no mention of the role of the chordae tendinae is preventing the AV valves turning inside out. Candidate B scores 5 marks.

Diastole: When the atria are relaxed blood is returned via the venae cavae (right atrium) and the pulmonary veins (left atrium) ✓. As blood returns, pressure increases and, when the ventricles relax, pressure in the atria increases to a point above that in the ventricles so that the atrioventricular valves are pushed open ✓. Both ventricles then fill passively ✓.

Overall, this is a good account even if some detail has been omitted. Candidate B earns 3 marks here giving a total of 11 marks for the biological content of the answer.

This is a well-structured account and ideas are expressed fluently. The links between electrical activity and contraction, and between pressure changes and valve action, are made clearly. The candidate scores 2 marks for the quality of written communication.

Overall, Candidate A scores 7 marks and Candidate B scores 13.

Section B total for Candidate A: 7 marks out of 15

Section B total for Candidate B: 13 marks out of 15

Paper total for Candidate A: 40 marks out of 75

Paper total for Candidate B: 64 marks out of 75